Cu/SBA-15 model catalysts for methanol steam reforming

A study on structure of oxidic precursors, catalysts, and on functionality of catalysts

vorgelegt von

Diplom-Chemiker

Gregor Koch

aus Berlin

von der Fakultät II - Mathematik und Naturwissenschaften

der Technischen Universität Berlin

zur Erlangung des akademischen Grades

Doktor der Naturwissenschaften

- Dr. rer. nat. -

genehmigte Dissertation

Promotionsausschuss:

Vorsitzender: Prof. Dr. Thomas Friedrich, TU Berlin

1. Berichter: Prof. Dr. Thorsten Ressler, TU Berlin

2. Berichter: Prof. Dr. Malte Behrens, Universität Duisburg-Essen

Tag der wissenschaftlichen Aussprache: 6. Juli 2015

Berlin 2015

Bibliographic information published by the Deutsche Nationalbibliothek

The Deutsche Nationalbibliothek lists this publication in the Deutsche Nationalbibliografie; detailed bibliographic data are available on the Internet at http://dnb.d-nb.de .

Dissertation, Technische Universität Berlin, 2015

ISBN 978-3-8325-4131-6

Logos Verlag Berlin GmbH
Comeniushof, Gubener Str. 47,
10243 Berlin
Tel.: +49 (0)30 42 85 10 90
Fax: +49 (0)30 42 85 10 92
INTERNET: http://www.logos-verlag.de

“. . . *for every chemical you trade a piece of your soul* . . .”

Billy Corgan in *This Time*

Zusammenfassung

Cu-Nanopartikel wurden auf das mesoporöse Silikat (SBA-15) aufgebracht, das eine große Oberfläche aufweist. Die synthetisierten Cu/SBA-15-Katalysatoren erwiesen sich als geeignete Modelkatalysatoren für die Methanol-Dampf-Reformierung. Eine Reihe von verschiedenen Cu/SBA-15-Katalysatoren mit Cu-Beladungen zwischen 10.9 und 27.1 Massen% wurde hergestellt. Es wurde Cu-Citrat als Cu-Präkursor verwendet und die Pulver wurden entweder als *dünne Schicht* oder als *dicke Schicht* kalziniert. Daraus ergaben sich zwei verschiedene Arten oxydischer Präkursoren, die jeweils die gleiche Cu-Beladung aufwiesen (XRF). Die Charakterisierung der Katalysatorstrukturen erfolgte mit Hilfe von N_2-Physisorption, DR-UV/Vis-Spektroskopie, XRD und XAS. *Dünn-Schicht-Präkursoren* hatten fein verteilte und amorphe CuO_x Teilchen, während *Dick-Schicht-Präkursoren* mehr geordnete und größere CuO_x-Teilchen hatten. Die strukturelle Unordnung in den Cu-Teilchen war scheinbar abhängig von deren Größe. Die Mesoporenstruktur des Trägers blieb erhalten. TPR-Messungen während der Aktivierung der Katalysatoren zeigten unterschiedliche Reduzierbarkeiten der CuO_x-Teilchen in Abhängigkeit von der Kalzinierungsmethode beziehungsweise der Cu-Beladung.

Mit Hilfe von *in situ* XAS und *in situ* XRD wurden nach der Aktivierung metallische Cu-Teilchen auf dem Träger gefunden. Die aktivierten *Dünn-Schicht-Katalysatoren* bestanden aus kleineren und wenig geordneten Cu-Nanopartikeln. Diese glichen sich in allen *Dünn-Schicht-Katalysatoren* obwohl die Cu-Beladung variierte. *Dick-Schicht-Katalysatoren* bestanden aus strukturell geordneten Cu-Nanopartikeln, die mit zunehmender Cu-Beladung größer wurden. Die mit N_2O gemessenen Kupferoberflächen stimmen mit diesen Ergebnissen überein. In der Methanol-Dampf-Reformierung zeigten die *Dünn-Schicht-Katalysatoren* größere H_2-Bildungsraten und größere H_2-Produktion je Oberflächenkupferatom im Vergleich zu entsprechenden *Dick-Schicht-Katalysatoren*. Die katalytische Aktivität korrelierte sowohl mit der Cu-Teilchen-Größe als auch mit dem Maß an struktureller Unordnung in den Cu-Teilchen.

Die H_2-Bildungsraten von allen Cu/SBA-15-Katalysatoren konnte gesteigert werden. Dies gelang entweder durch temporäre Zugabe von Sauerstoff zum Methanol-Dampf-Reformierungs-Edukt-Gemisch oder durch Redox-Aktivierung. Zusätzlich zeigten die *Dick-Schicht-Katalysatoren* größere intrinsische Aktivitäten. Die teilweise oder vollständig oxidierten Cu-Teilchen wurden wieder reduziert, was aber nicht zu Cu-Teilchenwachstum führte. *In situ* XRD und *in situ* XAS Analysen deuteten darauf hin, dass die strukturelle Ordnung in den Cu-Teilchen nach der Redox-Aktivierung größer war. Nach der Redox-Aktivierung wurden auch größer Kupferoberflächen gemessen.

Schlagwörter: Cu-Nanopartikel, nanostrukturiertes Silikat, SBA-15, *in situ* XRD, *in situ* XAS, TPR, Cu-Oberfläche, Methanol-Dampf-Reformierung, Redox-Aktivierung. Struktur-Aktivitäts-Korrelationen

Abstract

Cu nanoparticles were deposited on high surface area mesoporous silica SBA-15. The resulting Cu/SBA-15 catalysts were suitable model catalysts in methanol steam reforming. A set of various Cu/SBA-15 catalysts was prepared exhibiting Cu contents between 10.9 and 27.1 weight per cent (XRF). Using Cu citrate as Cu source, calcination as *thick layer* or *thin layer* resulted in two different types of oxidic precursors for catalysts exhibiting the same Cu loading. Structural characterization was achieved employing N_2 physisorption, DR-UV/Vis spectroscopy, XRD, and XAS. *Thin layer precursors* possessed well dispersed amorphous CuO_x particles, while *thick layer precursors* possessed more ordered and less dispersed CuO_x particles. The disorder in Cu metal particles was apparently size related. The mesoporous structure of the support was retained. TPR measurements during activation of catalysts revealed various reducibilities of CuO_x particles depending on calcination mode and Cu loading.

Various Cu metal particles present on SBA-15 were found by using *in situ* XRD and *in situ* XAS for structural characterization after activation and during methanol steam reforming. The activated *thin layer catalysts* consisted of smaller and more disordered Cu metal particles. Independent of Cu loading similar Cu metal particles were found (2 nm). *Thick layer catalysts* exhibited more ordered Cu nanoparticles increasing in size with increasing Cu content (2.5 nm - 8 nm). The Cu surface areas determined by N_2O chemisorption agreed with results from XRD and XAS. In methanol steam reforming at 250 °C, *thin layer catalysts* showed higher H_2 formation rates and higher H_2 TOFs compared to the corresponding *thick layer catalysts*. Catalytic activity correlated with both, degree of disorder and size of Cu metal particles.

Redox behavior of Cu/SBA-15 catalysts was tested by applying either temporary oxygen addition to methanol steam reforming feed or redox activation. After both procedures, increased H_2 formation rates of all Cu/SBA-15 catalysts were observed. Additionally, *thick layer catalysts* showed increased H_2 TOFs after redox activation. Partial or complete oxidation of Cu nanoparticles and subsequent re-reduction did not lead to increased sizes of Cu metal particles. *In situ* XRD and *in situ* XAS analysis indicated little increase in order of Cu metal particles after redox treatment. Larger Cu surface areas were also observed after redox treatment.

Keywords: Cu nanoparticles, nanostructured silica, SBA-15, *in situ* XRD, *in situ* XAS, TPR, Cu surface areas, methanol steam reforming, redox activation, structure activity correlations

Content

Abbreviations

CN	coordination number
CT	charge transfer
DESY	Deutsche Elektronen-Synchrotron
DR-UV/Vis	diffuse reflectance ultra violet and visible
DWF	Debye-Waller-Factor
e. g.	exempli gratia
et al.	et aliter
EXAFS	extended X-ray absorption fine structure
FWHM	full width half maximum
GHSV	gas hourly space velocity
HASYLAB	Hamburger Synchrotron Labor
i. e.	id est
LC	linear combination
LMCT	ligand to metal charge transfer
MeOH	methanol
MFC	mass flow controller
MPD	multichannel pixel detector
MSR	methanol steam reforming
NMR	nuclear magnetic resonance
pVf	pseudo-Voigt-function
QEXAFS	quick extended X-ray absorption fine structure
R	interatomic distance
SA	surface area
SAXS	small angle X-ray scattering
TCD	thermos couple device
TEM	transmission electron microscopy
TEOS	tetraethyl orthosilicate
TOF	turn over frequency
TOS	time on stream
TPD	temperature programmed desorption
TPO	temperature programmed reduction
TPR	temperature programmed reduction
XANES	X-ray absorption near edge structure
XAS	X-ray absorption spectroscopy
XRD	X-ray diffraction
XRF	X-ray fluorescence

Introduction and experimental background

Introduction

Energy storage

The fast growing mankind consumes more and more energy. From 1990 to 2008 the global population grew by 27 %. Simultaneously, the energy consumption per person increased globally by 10 % resulting in an absolute increase of energy requirement of 37 %. Energy from coal or oil contributes strongly to CO_2 release and therefore to the greenhouse effect threatening the stability of climate. Other solution must be found to prevent a climate collapse while sufficient energy supply is ensured. Besides nuclear power plants, alternative energy sources such as photovoltaics and wind turbines become more important. However, there will be still the problem of continuous energy production. Hence, energy must be stored. Storage devices for electrical energy are accumulators, being expensive and limited in capacity. Alternatively, energy can be stored in chemical compounds. In particular, hydrogen has been discussed as storage media and energy carrier. Since hydrogen fuel cells work sufficiently to produce electricity with adequate efficiency factors hydrogen became more important. However, storing hydrogen is still a challenge. Metal hydrides are too heavy for mobile applications. Porous metal organic frameworks are mostly instable. Compressing hydrogen in tanks reduces the efficiency factor and is thus not favorable.

Hydrogen storage may thus be managed by transforming it into another chemical for instance methanol (H_3COH). Methanol is nowadays a key chemical exceeding annual production of approximately 50 million tons. It is mainly produced by conversion of synthesis gas (H_2, CO, CO_2) obtained from natural gas. Minor source for methanol is fermentation by microorganisms.[1] Moreover, researchers are looking towards direct hydration of CO_2 to produce methanol.[2,3] Using CO_2 removed from the atmosphere and using H_2 obtained from electrolysis with renewable energies, methanol would be a green H_2 storage. Advantage of methanol compared to H_2 lay in its handling, due to liquid state. Therefore, as promising energy carrier, methanol can be distributed using existing infrastructure of ships, tanks, fuel stations etc. Methanol must later release previously bonded H_2, which can be subsequently used for supplying fuel cells. At this point methanol steam reforming becomes interesting.

Methanol steam reforming

The reaction of methanol with water is denoted as methanol steam reforming. It proceeds ideally to hydrogen and CO_2 employing catalysts at 250 °C.[4,5] This would be a CO_2 neutral process, if CO_2 for methanol synthesis was removed from atmosphere. There are two types of methanol steam reforming catalysts: One type bases on palladium metal and the other type bases on Cu metal as necessary components.[5] Besides of economic reasons the Cu catalysts are interesting as methanol steam reforming catalysts due to its frequent use in methanol synthesis.[6–8] According to *principle of microscopic reversibility*[9] Cu catalysts are also active in methanol steam reforming[10–12]. Since 1980s the industrial $Cu/ZnO/Al_2O_3$ catalysts were objects of investigation to reveal the relationships between structural characteristics and functionality primarily for methanol synthesis[7] and later for methanol steam reforming[10,11,13]. It was found that Cu metal particles are the prerequisite for activity in methanol steam reforming.[5,14] ZnO and Al_2O_3 are discussed as promotors for Cu metal particles. The influence of each component on the structure of Cu metal particles is still under debate.[2,6,8,12,15,16] In principal, catalytic activity is affected by chemical composition, preparation process, aging, calcination and activation.[5,17,18] The interaction of chemical compounds in each step may influence the structural characteristics of the activated catalysts leading to different activities. Therefore, the complexity must be reduced in order to disentangle the contribution of each synthesis parameter. The concept of model catalyst needs to be used.

For example, $Cu/ZnO/Al_2O_3$ catalysts were investigated varying the preparation route. Furthermore, more simple Cu/ZnO model catalysts were investigated varying chemical composition[19], preparation route[8,20,21] and the pretreatment[22]. Interestingly, the intrinsic activity of Cu metal particles was not the same for the catalysts of varying composition and varying pretreatment. Therefore, not only the dispersion of Cu metal particles but also the structure of Cu metal particles is crucial for catalysts performance.

Especially, the microstrain in Cu metal particles of Cu/ZnO and $Cu/ZnO/Al_2O_3$ catalysts leads to increased methanol steam reforming activity.[10,19] The origin of microstrain could not be explained in detail. ZnO is associated with microstrain in Cu metal particles. Microstrain was not observed in Cu metal particles deposited on ZrO_2[23] or SiO_2[16] as support. Moreover, small nanoparticles are prone to possess disordered structures.[24] Thus, the relationship between microstrain in Cu metal particles, increased methanol steam reforming activity and ZnO as promotor is still under debate. Additionally, structural variation of Cu metal particles accompanied by variation of catalytic activity can be evoked by oxygen co-feeding.[11,23,25] With the objective to elucidate correlation between microstructure of Cu metal particles and methanol steam reforming activity more simple model catalysts are required.

In this work Cu/SBA-15 model catalysts were chosen for detailed studies. The mesoporous silica, SBA-15, provides large surface areas and uniform pores. Furthermore SBA-15 is inactive in methanol steam reforming and stable under applied conditions. As support material, SBA-15 stabilizes Cu metal particles[25,26] and promise thus new insights in structure activity correlation by excluding ZnO. Well dispersed Cu metal particles on the support can be prepared using Cu citrate as precursor.[27,28] By varying the Cu loading and one synthesis parameter, while other parameters were kept invariant, influences of both might be distinguished. Therefore, systematic analysis of Cu metal particle structure and the catalytic activity in methanol steam reforming were the reliable basis for discussing structure activity correlation found in most simple Cu catalysts for H_2 generation by methanol steam reforming.

Outline of this work

Cu/SBA-15 catalysts obtained from citrate route were chosen as model catalysts. The preparation of various Cu metal particles was achieved by varying calcination mode and by varying the Cu loading. Prerequisite was the exclusion of ZnO being a strong promotor for Cu catalysts. It was the aim of this work to elucidate correlation between structure of Cu metal particles and their catalytic activity in methanol steam reforming. Therefore, complementary techniques and *in situ* technics as well as surface measurements were applied to reveal the structural differences of Cu metal particles.

First, the synthesis, structural and functional characterization of oxidic precursors was performed. Various CuO_x/SBA-15 catalysts were synthesized and the structure of CuO_x particles was deduced from results of DR-UV/Vis, XRD and XAS measurements.

Second, the oxidic precursors were activated to Cu/SBA-15. The freshly formed Cu metal particles were characterized with respect to particle size, microstructure, and their Cu surface areas. The activated Cu/SBA-15 catalysts were tested in methanol steam reforming. Catalysts performance was correlated to structural characteristics.

Third, temporary co-feeding of oxygen yield enhanced activity. Few structural changes were observed. To disentangle the influence of microstructure of Cu metal particles and the dispersion of Cu metal particles, a redox cycle was performed prior to activation of Cu/SBA-15 catalysts, simulating the oxygen co-feeding. The redox activated catalysts were investigated with regard to catalytic activity, Cu surface area, and structural changes of intermediate CuO_x particles as well as Cu metal particles. The found enhanced catalytic activity was correlated to surface areas of Cu metal particles and structural changes.

In this work the synthesis of various types of CuO_x particles supported on SBA-15 was shown. These various CuO_x particles were transformed to Cu metal particles preserving the structural differences. Employing the redox-activation to these Cu/SBA-15 catalysts led to catalysts showing further various activities and few structural changes.

These findings are discussed and compared with previously reported concepts such as *chemical memory effect* and microstrain-activity-correlation derived from other model catalysts.

Experimental background

Mass spectrometry *(MS)*

Process monitoring requests data measured at frequent intervals. A mass spectrometer is able to record the evolution of qualitative gas phase composition by measuring at frequent intervals. The compounds of continuously injected reaction gas are ionized by an electron beam. According to various masses of the molecules various ions are formed exhibiting various mass to charge ratios. Due to fragmentation and higher ionization states, one molecule is divided into several ions. These ions are separated and monitored as ion currents. The highest mass to charge ratio identifies the molecular weight of a certain compound. Ions of several compounds and ions of fragments often contribute to the same ion current. Therefore, quantification of compounds is difficult. However, changes in the reaction product distribution are accessible in good time resolution.

Gas chromatography (GC)

Gas chromatography enables separating vaporable substances, identifying them, and quantifying them. The reaction gas is injected into carrier gas using injection loops for automatic process monitoring. The remained reactants and products of methanol steam reforming are either gases or substances exhibiting high vapor pressure under the here employed conditions at 250 °C. Additionally, partial pressures were reduced by diluting the feed. The gas mixture is separated due to various interactions between the compounds of interest and the stationary phase. Here, a molecular thieve is used as stationary phase separating He, Ar, H_2, N_2, O_2 and CO. Furthermore, another stationary phase (polysilaoxane) separates CO_2, H_2O, methanol, formaldehyde, and methyl formate. Chromatograms are measured using a thermal conductivity detector.[29]

X-rays sources

X-rays are electromagnetic radiations exhibiting wavelength in the range of 0.01 nm to 10 nm. X-ray sources are X-ray tubes for laboratory applications or synchrotrons for more intense radiations. Using synchrotron X-rays, it is possible to use monochromatic radiation over a wide range of desired wavelength by employing monochromators. Moreover, the wavelength of the incident beam can be varied quickly. Therefore, synchrotron X-ray beams are commonly used in X-ray absorption spectroscopy. For X-ray diffraction, constant monochromatic radiation is sufficient. Hence, employing X-ray diffraction experiments is possible in single machines installed in laboratories.[24]

X-ray diffraction (XRD)

The electromagnetic radiation interacts with the electrons of an atom. The atoms create a regular lattice in crystalline solids exhibiting long range order. Due to similar wavelengths of X-rays and interatomic distances, the X-ray beam is diffracted at the electrons of atoms arranged regularly in the lattice. Constructive interferences arise at certain diffraction angles. Hence, crystalline solids show characteristic diffraction pattern. The number of observed diffraction peaks, *i.e.* positive interferences of diffracted beams, depends on the type of lattice. The lattice parameters decide on the diffraction angles, where the diffraction peaks are measured. Additionally, the intensity of diffraction peaks is governed by atom types, *i. e.* number of electrons.[24,30]

The shape of diffraction peak provides additional information about the microstructure of solids. Besides isotropic stretching or contracting of the crystals expressed by peak position shifts, the degree of anisotropic distortion, named microstrain, can also be estimated. Point defects, impurities, inclusions, epitaxial growth, and little shifts of atom layers cause microstrain. The microstrain evokes a diffraction peak profile showing additional contributions of Gaussian function to the typical Lorentzian function.[24,30]

X-ray fluorescence spectroscopy

The energy of X-rays radiated by a X-ray tube is sufficient to remove inner core electrons of atoms. The created hole is then re-occupied by an electron dropping from orbitals of higher energy. During this relaxation, radiation of characteristic energy is emitted corresponding to a certain transition in a specific element atom. The characteristic energy of radiation depends on type of atom and can therefore be used for elemental analysis. On basis of intensities of observed element specific peaks in the fluorescence spectra, the element fractions can be calculated. However, overlapping of signals and amount of light elements (lighter than fluorine in the periodic table) must be considered to receive reliable results.[31]

X-ray absorption spectroscopy (XAS)

Incident X-ray beams on a sample can be absorbed yielding excitation of inner orbital electrons to unoccupied orbitals or to the continuum. X-rays are absorbed when the energy of incident beam corresponds to the binding energies of inner orbital electrons. Then a sharp edge is observed at the specific energy in a XAS spectrum. The specific edges in XAS spectra are divided into two ranges: XANES (X-ray absorption near edge structure) and EXAFS (extended X-ray absorption fine structure). XANES is assigned to the near edge region including the absorption edge and approximately further 50 to 200 eV. EXAFS region lays between 100 eV to approximately 1000 eV behind absorption edge energy.[32]

X-ray absorption near edge spectroscopy (XANES)

At the absorption edge, the inner orbital electrons are excited to unoccupied orbitals and with increasing absorption energy to the continuum. The XANES gives information about the chemical state of an absorber atom. This comprises the electronic structure of an absorber and the coordination sphere. Furthermore, the position of the absorption edge is influenced by the oxidation state of the absorber atom. Increased oxidation states lead to enhanced electrostatic attraction of electrons to the core and thus to a shift to higher energies of the absorption edge. For example, the oxidation state of V atoms can be determined from the energy position of half edge jump height.[33] Oxidation states of Mo atoms are determined from energy position of a XANES feature in post-edge region.[34] With the objective to determine the oxidation state of Cu atoms one feature in the Cu K edge XANES was analyzed (dashed line in Figure 0-1). The energy position of the maximum in the XANES between 8.993 keV and 8.998 keV exhibited a good linear correlation as illustrated in Figure 0-1.

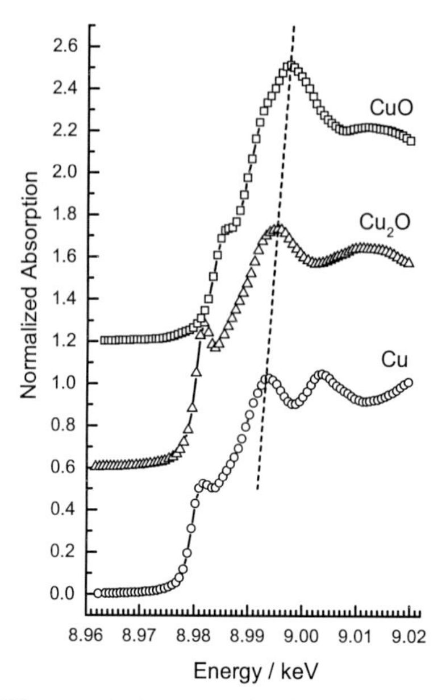

Figure 0-1 Normalized XANES spectra of reference materials Cu foil (circles), Cu_2O (triangles), and CuO (squares). Dashed line illustrates feature for determination of average oxidation state of Cu atoms.

Reference bulk materials and mixture of reference materials were used as calibrations standards (Figure 0-2). Since the mixtures of reference compounds were in good agreement, shifts and overlapping effects in data of binary mixtures were estimated in the confidence band assuming certainty of 95 %. Confidential band is illustrated as dotted line in Figure 0-2.

Using the calibration function, shown in Figure 0-2, analysis of XANES spectra reveal the average oxidation state of Cu atoms in solids. Additionally, the particle size influences the energy position of this feature (compare *II.3.4*) and must be considered. The average oxidation state can also be derived from the fractions of Cu or copper oxides in a sample. Reconstruction of XANES spectra with adequate reference spectra disentangles the phase contribution.[35] However, using linear combination of reference XANES spectra, the estimation of the average oxidation state of Cu particles was not adequate. The main problem was the difference in particle sizes of reference compounds and measured samples.[36] The reference spectra of crystalline materials did not represent the spectra of nanoparticles of the same material leading to large uncertainties (*II.3.3*).

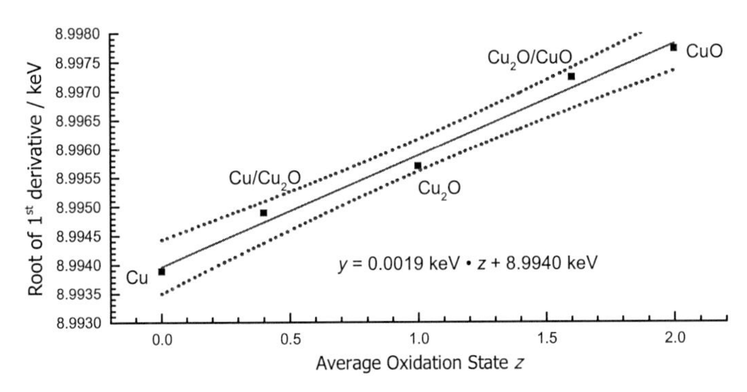

Figure 0-2 Correlation between average Cu oxidation state z and XANES maximum (see Figure 0-1). Crystalline references were measured as pure materials or as mixture. Dotted blue lines limit the confidence band with 95 % certainty.

Extended X-ray absorption fine spectroscopy (EXAFS)

At increasing energies being higher than the binding energy of an inner core electron of incident X-rays, the inner core electrons are excited to the continuum. The excited electron in the continuum is scattered at electrons of adjacent atoms. Due to interferences between outgoing and backscattering electrons, specific oscillation is detected depending on type, distance and the mean disorder of adjacent atoms. Hence, the EXAFS analysis provides information about the distance of atoms, the kind of atoms, and the static disorder in solids. These information are accessible by refining theoretical spectra to measured spectra. Therefore, a suitable model is needed. The basis of the model is often based on XANES analysis or other analytical data such as XRD.[37]

Diffuse reflectance ultraviolet and visual spectroscopy (DR-UV/Vis)

Electromagnetic radiation ranging from 200 to 850 nm is denoted as UV/Vis radiation. Radiation in this energy range can be absorbed by valence electrons of certain chemical compounds. The valence electrons contribute to chemical bonding. Hence, the energy absorbed by valence electrons depends on chemical bonding. Therefore, absorption spectra are affected by types of neighboring atoms. Furthermore, the absorption spectra reveal varying atom distances and show the oxidation state of a transition metal ion. It is possible to deduce the local environment of metal ions. Due to quantum confinement effect, particle sizes of particular metal oxides can be assessed from the energy position of the absorption edge.[17,38]

Solid samples are measured in diffuse reflectance mode. Kubelka-Munk-function is used to calculate the spectra. The following assumptions were made: absorbers are homogenously distributed, samples have to be infinitely thick (1-5 mm layer thickness), fluorescence does not occur, and reflectance is greater than absorption. The obtained Kubelka-Munk value should not excess two. Otherwise sample dilution is necessary for measuring reliable spectra.[38,39]

Determination of surface areas

The determination of surface areas bases on several physisorption and chemisorption methods depending on solids. Physisorption includes all adsorption processes of gases on solids exhibiting adsorption energies up to 80 kJ/mol. Adsorption processes exhibiting adsorption energies higher than 80 kJ/mol are denoted chemisorption. The adsorption and desorption processes are strongly depended on temperature. For reliable surface measurements temperature-programmed reduction (TPR), temperature-programmed oxidation (TPO) and temperature-programmed desorption (TPD) are needed. In a first step detectable particles have to be formed or the surfaces have to be cleaned from other adsorbed species. The specific preparation of samples enables quantifying adsorbed molecules and thus the number of possible active sites.[40]

N_2 physisorption

N_2 physisorption isotherms give information about the porosity of solids and reveal their entire surface area. Furthermore, N_2 physisorption isotherms can elucidate the fraction of micropores (0 - 2 nm) and the size distribution of mesopores (2 - 20 nm). It is assumed that N_2 molecules cover first a monolayer on the surface and then N_2 molecules form multilayers on the surface. Moreover, interactions between the N_2 molecules and interactions between N_2 molecules and solids are neglected. The entire surface areas can be calculated using the BET equation.[41] The adsorbed N_2 multilayers grow condensing finally in mesopores. According to the BJH method basing on the Kelvin-equation[42], the average mesopore diameters, the average mesopore surface areas, and the distribution of mesopore size can be estimated.

Temperature-programmed reduction (TPR)

The consumption of H_2 during temperature-programmed reduction depends on the amount of reducible species and on the corresponding reducibilities (*i. e.* onset temperatures). The areas of the measured H_2 consumption peaks correlated directly to amount of oxygen transformed to H_2O. Hence, it can be used for quantification of corresponding reducible species. The H_2 consumption peaks may show various profiles such as broadened peaks, sharp peaks, peaks with shoulders, or multiple peaks. These features in peak profile enable distinguishing various intermediates during reduction or indicate a particle size distribution.[43]

Determination of Cu surface areas

N_2O decomposes quickly in presence of metallic Cu at 35 °C to N_2 and to oxide ion bonding on the Cu metal surface. The Cu surface is passivated, while two Cu surface atoms share one oxide ion. The temperature must be strictly controlled. Increased temperature leads to further oxidation of Cu atoms underneath the surface and is denoted as bulk oxidation. Accordingly, the number of Cu surface atoms is equally represented by either the consumption of N_2O, or by the amount of formed N_2, or by the amount of chemisorbed oxygen. Applying subsequently H_2 TPR, the amount of chemisorbed oxygen atoms can be determined. As result, corresponding number of Cu surface atoms can be calculated. Hence, the specific Cu surface areas can be determined based on Cu atom size and specific Cu surface plane.[44]

Chapter I Influence of calcination on structural configuration of CuO_x particles in CuO_x/SBA-15 catalysts

I.1 Introduction

Catalyst synthesis often proceeds in various steps. Mixing chemicals, precipitating solids at certain pH values, aging, drying, calcination, and activation have great impact on catalytic activity. A wide range of Cu based catalysts for methanol steam reforming has been examined[5], including variation of synthesis parameters and the corresponding effects on catalytic activity. For example, Behrens and Schlögl[18] gave an overview on synthesis parameters of Cu/ZnO catalysts and the related catalytic activity in methanol synthesis. Variation of more than one parameter during synthesis often leads to a more complex picture. Single correlations between parameters and activity of catalysts might not be easily accessible. The thorough analysis of precursors of Cu/ZnO catalyst revealed a *chemical memory effect*. For example, good distribution of Zn atoms substituting Cu atoms in the oxidic precursor with malachite structure resulted in well dispersed Cu particles in the active Cu/ZnO catalysts.[19] Thus, it is important, to study the structure of oxidic precursors of catalysts and to elucidate their structural differences. Then, a better understanding would be accessible between the effect of variation of a single synthesis step and later catalytic activity of catalysts.

In methanol steam reforming, the presence of Cu in catalysts is a prerequisite for well working catalysts. Therefore, the microstructure of Cu metal particles is of great interest. Previous investigation showed the influence of ZnO on the microstructure of Cu metal particles.[8,19,45,46] Therefore, ZnO must be omitted in the catalysts, to reveal correlation between microstructure of Cu particles and their performance in methanol steam reforming. Cu metal particles are often obtained by reduction of suitable copper oxide particles present in oxidic precursors. During reduction, the active catalyst is created. Consequently, well dispersed CuO particles lead to well dispersed Cu metal particles. As inhibitor of sintering, high surface silica are suitable support materials. Using support materials enables stabilization of Cu in a desired coordination sphere, and, in special cases, stabilizing certain oxidation states of copper atoms. Knowing the character of support materials thus enables assessing the stability of possible structures. After loading the support materials with copper precursors and subsequent calcination, the oxidic precursors are formed. The structure of copper oxide particles depends on the used support materials.[5] Besides zeolites, which partly provides Cu^+ embedded in special cage positions, silica based mesoporous materials are frequently used in catalysis research.[47] These materials differ in arrangement of pores, in pore diameter, and consequently in the available surface area.[48] CuO has been previously deposited on high surface silica gels[28] and mesoporous silica such as cubic MCM-41[49], and on SBA-15[26,50,51]. SBA-15 is a suitable support material, because the hexagonally arranged mesopores exhibiting 9 nm in diameter and the thick pore walls lead to high thermal stability.[52] Moreover, the two dimensional arrangement of mesopore channels reduces the mobility of supported metals. Consequently, diffusion based deactivations should be limited.[53]

Investigations of CuO supported on SBA-15 were carried out with regard to preparation route, *i. e.* linking CuO on support material: one pot synthesis[49], grafting of copper ions[54], applying dynamic vacuum during calcination[26], or varying the gas flows and gas compositions during calcination[50]. Grafting methods, application of dynamic vacuum or a gas flow of inert gas during calcination result in well dispersed copper oxide particles on the support. Hence, subsequent reduction leads to well dispersed Cu nanoparticles on the silica surface. It was reported that a diluted NO gas stream during calcination resulted in larger Cu metal particles than calcination under N_2 stream.[50] Another route to well dispersed metal oxides is the pyrolysis of metal citrates.[27] Looi *et al.*[28] and Suh *et al.*[49] showed for silica support and for SBA-15 support, respectively, an improved dispersion of copper oxide particles in the oxidic precursors by use of citric acid. The increased dispersion was assigned to the copper citrate complex, which separated copper ions and thus hindered crystal growth and agglomeration of CuO during calcination. Good separation of copper atoms in an oxidic precursors was also demonstrated by Drake *et al.*.[54] Increasing size of ligands resulted in well dispersed copper oxide particles after calcination. Oxidative decomposition of the precursors may result in isolated $[CuO_x]$ units, connected $[CuO_x]_n$ units such as clusters or chains, layers of $[CuO_x]_n$ units or CuO nanocrystals on the support material. The relationship between these various structural motives and their catalytic activity is still under debate.[55] Even as preliminary stage in catalyst synthesis, the various structural motifs in oxidic precursors may lead to Cu metal particles of various microstructures[56], particle shapes, or particle sizes[57].

In this chapter, synthesis routes of copper oxide particles and their characterization are presented. First, the effect of incipient wetness treatment with various chemicals on the pore structure of SBA-15 before calcination was investigated. Second, differences of CuO_x particles derived from pyrolysis of an ammonia copper citrate precursor linked on SBA-15 with regard to the CuO_x structure are discussed. Variation of CuO_x particles was achieved varying the synthesis conditions. First, the same calcination conditions were applied to SBA-15 with five different Cu loadings. Second, the mode of calcination was varied and applied to samples with various Cu loadings. Treated SBA-15 was arranged as *thin layer* or as *thick layer* in a crucible. The temperature program and the atmosphere (air) were kept invariant during calcination.

I.2 Experimental

I.2.1 Synthesis

Nanostructured silica support was synthesized according Zhao and coworkers.[58] 16.2 g of copolymer P123 (Aldrich) were dissolved in a mixture of 293.6 g H_2O and 8.8 g concentrated HCl. After stirring for 24 h at 35 °C, 32 g of TEOS (Aldrich) were added and stirred for 24 h. Subsequently, the solution was heated in a closed bottle for 24 h at 115 °C. The obtained white powder was separated and washed with ethanol, which was acidified with concentrated HCl. The white powder was dried at 105 °C for at least 2 h. Afterwards, the powder was calcined in air by heating to 180 °C at 1 °C/min and held for 3 h, and Subsequently, the powders were heated to 550 °C and held for 5 h at 550 °C.

Table I–1 Summary of amount of used chemicals for catalyst precursor synthesis. Cu content derived from XRF measurements. Number before *Cu* represents Cu loading as wt.%. * citric acid

sample	SBA-15 / g	copper citrate / g	NH_3 solution / ml	*thin layer precursors*	*thick layer precursors*
SBA_NH$_3$	0.25	-	0.35	-	-
SBA_Citrat	0.25	0.050*	0.35	-	-
2Cu	4.012	0.252	6.0	2.88 ± 0.07	-
10.9Cu	8.033	2.351	32	11.1 ± 0.2	10.9 ± 0.3
14.9Cu	8.004	3.735	32	14.74 ± 0.06	14.93 ± 0.07
16.3Cu	8.001	4.086	32	16.2 ± 0.4	16.3 ± 0.4
22.3Cu	8.000	5.836	31	23.01 ± 0.13	22.3 ± 0.2
27.1Cu	8.000	7.499	35	27.2 ± 0.2	27.1 ± 0.3

As preliminary study according to the incipient wetness method, freshly prepared SBA-15 was treated with ammonia (Merck, 25% *pro analysi*), citric acid (ROTH, ≥ 99.5%) dissolved in ammonia, and copper citrate (Pfaltz & Bauer, 98.9%) dissolved in ammonia. For catalyst synthesis, typically 8 g SBA-15 and varying contents of copper citrate dissolved in 32 ml ammonia were used (listed in Table I–1). In the following, Cu catalysts are denoted by the measured Cu content. Consequently, five catalysts are discussed: 10.9Cu, 14.9Cu, 16.3Cu, 22.3Cu and 27.1Cu. After drying in air, the catalyst precursors were divided into two fractions. One fraction was deposited as thin layer in a ceramic crucible in the muffle furnace. The other fraction of 10.9Cu, 14.9Cu, and 16.3Cu samples were deposited as thick layer in a ceramic crucible in the muffle furnace. The other fraction of 22.3Cu and 27.1Cu were calcined in a vertical silica tube furnace with a diameter of 30 mm and a layer thickness of about 2 mm.

After heating to 250 °C at 5 °C/min, the samples were held for 5 h at 250 °C. The resulting oxidic precursors consisted of copper oxide particles supported on SBA-15. The oxidic precursors obtained from the different calcination routes are called *thin layer precursor* and *thick layer precursor*.

I.2.2 N_2 physisorption

N_2 adsorption and desorption isotherms were measured at -196 °C using a BELSORP mini II apparatus. Prior to measurements, the sample holder, containing SBA-15 or oxidic precursors, were evaporated at 95 °C for 20 min. Subsequently, the temperature was raised to 175 °C and held for about 17 h. The surface of the materials was calculated according to BET method[41] in the range of $0.037 < p/p_0 < 0.25$ assuming an area of 0.162 nm^2 per N_2 molecule. Pore size distribution and mesopore surface area were calculated according to the BJH method.[42]

I.2.3 X-ray fluorescence analysis (XRF)

The copper content of the oxidic precursor was determined using X-ray fluorescence spectroscopy (XRF). 120 mg of sample and 120 mg of wax (Hoechst wax C micropowder, Merck) were mixed and pressed to a pellet of 13 mm in diameter. XRF spectra were measured with a PANalytical AXIOS spectrometer (2.4 kW) equipped with a Rh tube, a scintillation detector (E > 8500 eV) and a gas flow detector (500 eV < E < 8500 eV). Quantitative composition was calculated using the standardless analysis implemented in the SuperQ 5 software package (PANanlytikal).

I.2.4 X-ray diffraction (XRD)

X-ray diffraction patterns of oxidic precursors were measured on a PANalytical diffractometer (X'Pert PRO, MPD). Measurements were performed in reflection mode in θ/θ geometry using Cu K_α irradiation. The diffractometer was equipped with a solid state multichannel detector (PIXcel®). The range between 5 ° to 120 °2θ was scanned with a step size of 0.013 °2θ and measuring time of 0.5 s. Application of programmable slits provided an effective irradiation length of 12 mm/step.

I.2.5 Diffuse reflectance ultraviolet/visible-light spectroscopy (DR-UV/Vis)

Diffuse reflectance UV/Vis spectra were measured on a JASCO two beam spectrometer V670. The spectrometer was equipped with an integration sphere coated with barium sulfate. SBA-15 was measured as reflectance standard for baseline correction. Spectra were measured and subsequently truncated to a range of 5000 cm^{-1} to 40000 cm^{-1}. The measured absorbance was transformed to the Kubelk-Munk-function $F(R_\infty)$. Typically, catalyst powders were diluted with pure SBA-15, so that the Kubelka-Munk values remained in a range of $0 \leq F(R_\infty) \leq 2$. Profile fitting of UV/Vis spectra succeeded using four Gaussian functions. One was assigned to a d-d transition and three were assigned to charge transfer (CT) transitions. According to Scholz *et al.*[59], the full width at half maximum (FWHM) of CT transitions was correlated to be the same during the refinement procedure.

I.2.6 X-ray absorption spectroscopy (XAS)

XAS experiments were performed at beamline X at HASYLAB at DESY, Hamburg. Therefore, samples and BN were mixed and pressed with a force of 0.5 t for 10 s into pellets with 5 mm in diameter and a mass of 35 mg. Sample mass was calculated to obtain an edge jump of $\Delta\mu(d) = 1.2$. The samples were measured at the Cu K edge in the range from 8.900 keV to 9.940 keV in transmission modus, using a Si(111) double crystal monochromator. Data reduction and analysis were performed with the software package WinXAS.[60] Background subtraction and normalization were performed using a linear fit in the pre-edge range from 8.900 keV to 8.950 keV and a third–degree polynomial in the post-edge range between 9.105 keV and 9.930 keV of an absorption spectrum. Smooth atomic background $\mu_0(k)$ was calculated with cubic splines in the range 3.1 $Å^{-1} \leq k \leq 13$ $Å^{-1}$. Extracted $\chi(k)$ was k^3 weighted and then Fourier transformed into R space using a Bessel window. XAS fits were performed in the range from 1 to 3.2 Å of the pseudo radial distribution function. Detailed fit parameters and correlations are given in *I.3.10* and *I.3.11*.

I.3 Results and discussion

I.3.1 Composition of oxidic precursors

The effective Cu loading of oxidic precursors, CuO_x/SBA-15, was measured by XRF and the results are given in Table I–1. Besides silicon, oxygen and negligible amounts of zinc (Zn was found in traces in copper citrate) only copper was found. Furthermore, carbon was not detected by CHN-analysis. Thus, copper citrate decomposed completely.

I.3.2 Preliminary study on support stability using N_2 physisorption

SBA-15 support material showed a type IV physisorption isotherm characteristic for ordered mesoporous materials. Selected physisorption isotherms are depicted in Figure I-1, left and Figure I-2, on the top left. After treating SBA-15 according to the incipient wetness method and following calcination, the isotherms still showed type IV characteristics, but deviated from those of SBA-15. Especially the slope of the hysteresis loop was affected. The hysteresis loops of samples treated with ammonia or citric acid dissolved in ammonia, were slightly shifted to lower p/p_0 values and were lowered in height. Adsorption and desorption branches of hysteresis loops run parallel and tilted at p/p_0 values higher than 0.42 (see Figure I-1). Hence, mesopores remained open but were apparently contracted. However, applying Cu citrate dissolved in ammonia to SBA-15 resulted in an isotherm which is typical for mesoporous SBA-15. The mesopore diameter was invariant. Only the height of the hysteresis was reduced.

Table I–2 Results of N_2 physisorption isotherm analysis, according to BET-plot and BJH-plot (A_{pore} and d_{pore}). Pure SBA-15 was differently treated by variation of used solutions (second column). Calcination and amount of SBA-15 and solutions were kept constant.

sample	incipient wetness solution	A_{BET} / m^{2}g^{-1}	A_{pore} / m^{2}g^{-1}	d_{pore} / nm	A_{pore}/A_{BET}
SBA-15	-	731	644	9	0.88
SBA_NH$_3$	25 % NH_3	581	529	8	0.91
SBA_Citr	citric acid in 25 % NH_3	562	524	8	0.93
SBA_2.3Cu_Citr	copper citrate in 25 % NH_3	487	481	9	0.99

These changes, induced by citric acid, citric acid dissolved in ammonia, and copper citrate dissolved in ammonia, are pronounced in the BJH-plot in Figure I-1, right. The narrow pore size distribution of pure SBA-15 was broadened and shifted to smaller pore diameters after incipient wetness and calcination. Ammonia and citric acid dissolved in ammonia reduced the mean pore diameter form 9 to 8 nm. SBA-15 treated with these solutions also showed reduced entire BET surface areas and reduced pore surface areas (compare Table I–2). In contrast, copper citrate dissolved in ammonia prevented the mesopores from contracting. The major fraction of the mesopores showed the same diameter after calcination. Only a minor fraction of mesopores contracted slightly. However, the entire surface area was reduced. The ratio of mesopore surface area and entire surface area can be used as measure of the contribution of micropores of SBA-15 to the entire surface area. This ratio increased from 0.88 for pure SBA-15 to 0.91 for ammonia treated SBA-15 and 0.93 for citric acid dissolved in ammonia treated SBA-15 to 0.99 for 2Cu sample. Hence, SBA-15 was sensitive towards the basic character of ammonia solution and the citrates dissolved therein. Ammonia itself reduced the diameter of mesopores without loss of micropore surface area. Addition of citric acid to ammonia also reduced the diameter of the mesopores comparable to contraction evoked by ammonia, but additionally diminished the amount of micropores. Further addition of copper ions in form of Cu citrate dissolved in ammonia slightly affected the mesopore size. Then, micropores were not detected.

Similar observations has been reported for SiC formed in SBA-15.[61] It is likely that copper oxide particles were embedded in micropores. Micropores were gradually filled with increasing Cu content. Otherwise, the micropores were closed by an overlayer of copper oxide formed on the mesopore walls.

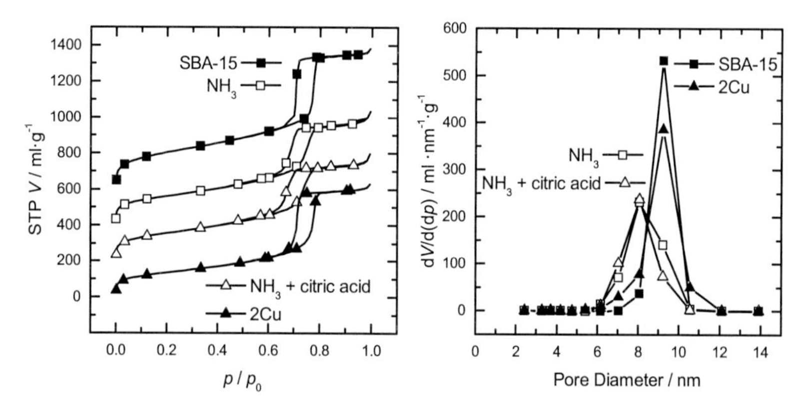

Figure I-1 N_2 physisorption isotherms on the left, stacked with an offset of 200 ml·g^{-1}, and BJH-plot of SBA-15 (filled squares) on the right, treated with ammonia (empty squares), treated with citric acid dissolved in ammonia (empty triangle), and treated with copper citrate dissolved in ammonia (filled triangle).

I.3.3 Mesoporous structure of oxidic precursors

N_2 physisorption isotherms of pure SBA-15 and oxidic precursors are shown in Figure I-2, left. A type IV hysteresis that is characteristic for materials with uniform mesopores was found for all oxidic precursors. The first incline up to $p/p_0 = 0.03$ resulted due to filling of intrawall micropores.[62] The largest amount of micropores was found in SBA-15. The entire surface areas of here investigated SBA-15 were similar to those reported.[58] The standard deviation of BET surface areas fresh SBA-15 was about 2.3 %. Accompanied by nearly uniform mesopores (BJH-plot) having a diameter of 9 nm, the different SBA-15 powders were suitable starting materials for catalyst synthesis. The reported influence of support surface area on sizes of CuO_x particles[63] can be neglected in further considerations due to homogeneity of used SBA-15.

Table I–3 Comparison of BET surface areas and mesopore surface area contributions to the entire surface areas of SBA-15, *thin layer precursors*, and *thick layer precursors*.

	pure SBA-15		oxidic precursors			
			thin layer precursors		*thick layer precursors*	
sample	A_{BET} /m^{2}g^{-1}	A_{pore}/A_{BET}	A_{BET} /m^{2}g^{-1}	A_{pore}/A_{BET}	A_{BET} /m^{2}g^{-1}	A_{pore}/A_{BET}
10.9Cu	742	0.87	393	1.00	434	0.98
14.9Cu	716	0.89	362	1.02	411	0.98
16.3Cu	763	0.90	344	1.03	379	0.98
22.3Cu	736	0.89	343	1.03	369	0.98
27.1Cu	759	0.87	308	0.99	342	0.97

After deposition of copper precursors and their oxidative decomposition to CuO_x/SBA-15 the character of pores changed distinctly depending on calcination mode. For oxidic precursors a significant loss in surface area was found according to higher densities of these materials due to deposition of copper oxide particles. All oxidic precursors showed a decreased contribution of micropores to the surface area compared to SBA-15. The micropores could be filled or closed by deposited CuO_x particles (see preliminary study). An often discussed decrease of surface area of SBA-15 due to restructuring of pore walls and intrawall micropores during repeated calcinations and hydrothermal treatments[52] can be neglected. Since the reported conditions were not reached here (650 °C during calcination or 30 % H_2O in N_2 at 400 °C during hydrothermal testing), the Cu citrate precursor or the decomposition of this precursor apparently smoothed the pore walls. Although, additions of etching ammonia or citric acid dissolved in ammonia led to reduction of mesopore diameter. CuO_x particles stabilized the pore walls and therefore, no contraction of mesopores was observed.

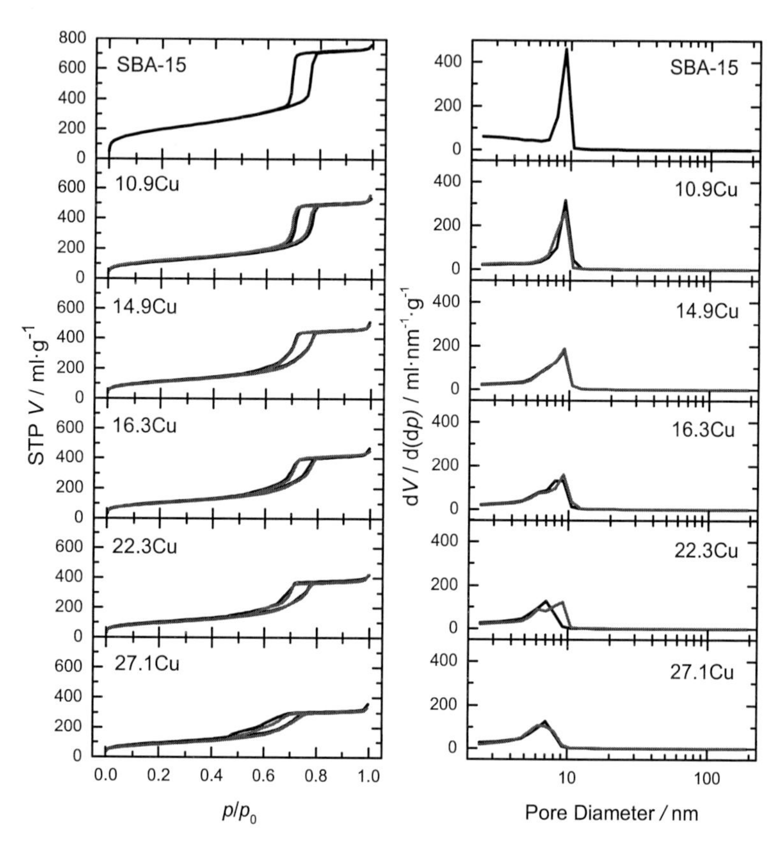

Figure I-2 Left: N_2 physisorption isotherms for exemplary SBA-15 on the top and the oxidic precursors below. Right: Comparison of BJH-plots of exemplary SBA-15 on the top and the oxidic precursors below. *Thick layer precursors* are depicted in black, and *thin layer precursors* are depicted in green. Cu loadings are given as number in front of *Cu*.

According to Figure I-2 the shape of the hysteresis loop at $0.45 < p/p_0 < 0.8$ changed with increasing Cu loading. The steep incline of the adsorption branch and the parallel decline of the desorption branch observed at SBA-15 indicated uniform pores with a narrow pore size distribution. With increasing Cu loading, this slope decreased (Figure I-2, left). The corresponding broadening of the pore size distribution is shown in the BJH-plot (Figure I-2, right). The decrease of pore diameter was caused by CuO_x particles deposited in mesopores.

Since desorption branches converged with the adsorption branches at p/p_0 higher than 0.42, blocking of mesopores was excluded.

The surface area of the mesopores can be calculated using the BJH method.[42] Surface areas of mesopores are summarized in Table I–3. The ratio of the mesopore surface area and the entire surface area for SBA-15 was between 0.87 and 0.90. While the *thin layer precursor* showed a ratio between 0.99 and 1.03, the *thick layer precursor* possessed a ratio between 0.97 and 0.98. Both, micropores and mesopores contribute to the entire surface. In *thin layer precursors*, mainly mesopores contributed to the entire surface area. *Thick layer precursors* revealed a higher contribution of micropore surface area to the entire surface area compared to *thin layer precursors*. Additionally, larger BET surface areas in *thick layer precursors* than in *thin layer precursors* indicated the presence of more micropores in *thick layer precursors*. Hence, copper oxide was deposited in the micropores as well as in the mesopores, while the mesoporous structure of silica support was retained. Furthermore, in *thin layer precursors*, more micropores were filled with copper oxide than in *thick layer precursors*.

I.3.4 Cu atom density on silica surface

The density of Cu atoms on the available silica surface area after calcination was calculated. Results are shown in Figure I-3. The Cu atom densities of *thin layer precursors* (green triangles) were higher than that of *thick layer precursors* (black squares). Both Cu atom densities were linearly correlated to the amount of copper oxide present on SBA-15. The higher Cu atom densities of *thin layer precursors* may be caused by different shapes of copper oxide particles present in *thin layer precursors* and *thick layer precursors*. Apparently, the available silica surface changed according to the applied calcination route. In *thin layer precursors* copper oxide particles were likely located in the micropores and therefore limited in growth by pore walls. Larger copper oxide particles of *thick layer precursors* (compare XRD *I.3.5*, UV/Vis *I.3.7*, XAS *I.3.11*) were mainly formed in the mesopores. Thus not enough copper oxide was available in *thick layer precursors* to fill the micropores and to form larger copper oxide particles in mesopores.

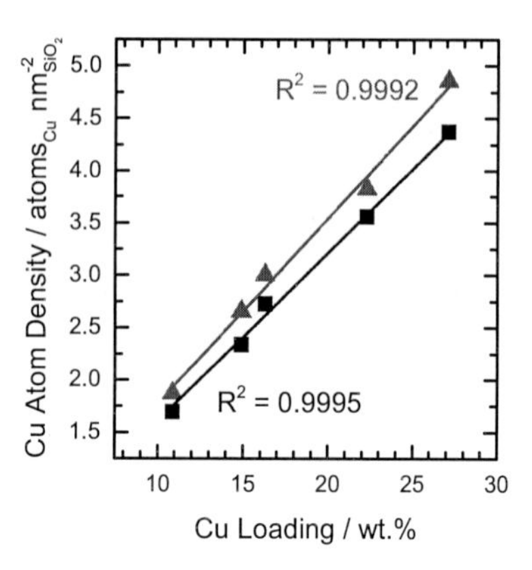

Figure I-3 Cu atom density on SBA-15 surface of *thin layer precursors* (green triangles) and *thick layer precursors* (black squares).

Mesopore blocking can be neglected due to the good correlation between calculated Cu atom density and copper loading. Blocked mesopores would decrease the observed silica surface area. In case of pore blocking, there would be expected a deviation to higher Cu atom densities for higher Cu loadings. Therefore, the good linearity was a further indication for entire preservation of mesopore structure.

The degree of connectivity of metal oxide units can be assessed by surface coverage of metal oxide units according to reference.[59] The surface area occupied by a [CuO_6] unit was estimated to be 0.24 nm^2 (circle with diameter of axial O-Cu-O bond (Figure I-10)). For the lowest Cu atom density of 1.8 at least two [CuO_6] units may be linked.[59] For a Cu atom density of four and higher, a monolayer of linked [CuO_6] units can be assumed. Otherwise, agglomerated [CuO_x] units formed larger CuO_x particles, which were well distributed over the silica surface.

I.3.5 X-ray diffraction of oxidic precursors

X-ray diffraction patterns of oxidic precursors are depicted in Figure I-4. All catalysts showed a broad peak at 24 °2θ, which is caused by amorphous silica of SBA-15.[64] XRD patterns of *thin layer precursors* (green, top) showed no further peaks indicating amorphous or very small copper oxide particles supported on SBA-15. Conversely, XRD patterns of *thick layer precursors* (black, bottom), with a loading higher than 10.9 wt. % showed small and broad peaks around 35 °2θ, which corresponded to CuO.[65] Additionally, for 27.1Cu *thick layer precursor* contribution of Cu_2O[66] was found. *Thick layer precursors* with loadings of more than 14.9 wt.% Cu showed small diffraction peaks indicating nanoparticles of CuO present on SBA-15 support.

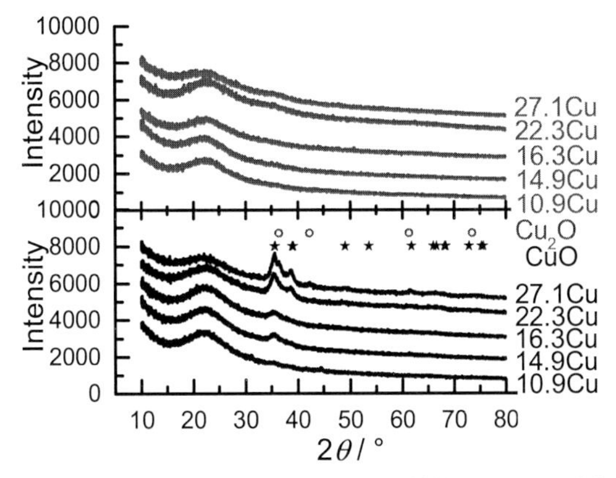

Figure I-4 XRD patterns of *thin layer precursors* on the top (green), *thick layer precursors* below (black), and significant diffraction peak position of crystalline CuO (black stars) and Cu_2O (red circles). Measured data are depicted with an offset.

I.3.6 DR-UV/Vis of oxidic precursors

The freshly calcined powders of oxidic precursors differed in color. While *thick layer precursors* were brown powders *thin layer precursors* were green powders. Therefore, DR-UV/Vis spectroscopy was a suitable method to distinguish various structures of copper oxides present on SBA-15. The recorded DR-UV/Vis spectra of 10.9Cu and 27.1Cu *thin layer precursor* and fitted profile functions are depicted in Figure I-5. Four different absorption bands were necessary to reconstruct the DR-UV/Vis spectra. Single absorption bands were overlapping in the DR-UV/Vis spectra. It is therefore difficult to distinguish various types of $[CuO_x]_n$ units and their quantity present on SBA-15 with DR-UV/Vis spectroscopy. However, structural difference can clearly be seen.

Absorption bands were located around 14000 cm^{-1}, between 21500 cm^{-1} and 25000 cm^{-1}, from 30000 cm^{-1} to 33000 cm^{-1} and around 41000 cm^{-1}. The position of the band maximum and the relative contribution to the overall absorption varied and indicated differences in $[CuO_x]_n$ structures present on SBA-15.

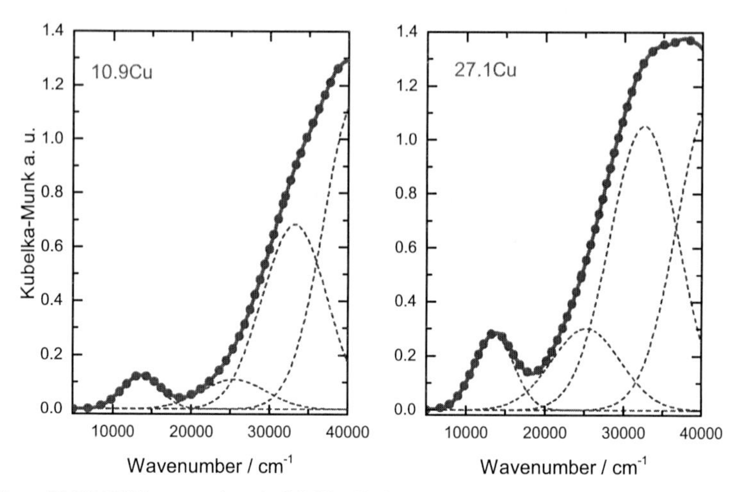

Figure I-5 DR-UV/Vis spectra (green) of 10.9Cu *thin layer precursor* left and 27.1Cu *thin layer precursor* right. Blue broken lines represent fitted Gaussian functions of single contributions of each absorption band. The fitted spectra (blue dots) overlap nearly ideally with measured spectra (green).

In the range of 12500-16666 cm^{-1}, d-d transition of Cu^{2+} in an octahedral coordination sphere occurs.[67] Cu^{2+} exhibits an electron configuration of $[Ar]3d^9 4s^0$. Thus, polyhedrons possessing a Cu^{2+} center exhibit a Jahn-Teller-distortion.[68] As consequence, $[CuO_6]$ octahedrons show d-d absorption bands at 12500 cm^{-1}. This absorption band shifts with increasing tetragonal distortion to 15400 cm^{-1} for crystalline CuO[67] up to 16666 cm^{-1} for nearly square planar Cu^{2+} coordination spheres[63]. The d-d transitions of all *thin layer precursors* were found around 13590 cm^{-1}. This indicated the presence of similar distorted $[CuO_6]$ octahedrons independent of Cu loading. These distorted $[CuO_6]$ octahedrons, present on SBA-15, resembled the copper coordination sphere in the copper silicate dioptase (13300 cm^{-1})[69], where two Cu^{2+} ions are connected by two O^{2-} ions of two different $[SiO_4]$ tetrahedron (see middle of Figure I-14). In comparison, the d-d transition absorption band of the *thick layer precursors* shifted from 13700 cm^{-1} for 10.9Cu up to 14900 cm^{-1} for 27.1Cu with increasing copper loading. This hypsochromic effect indicated a shortening of the Cu-O distance[63] and, thus, an increasing distortion of $[CuO_6]$ units. Hence, with increasing Cu content the coordination sphere of Cu^{2+} present in *thick layer precursors* resembled that of bulk CuO, consistent with results from XRD measurements (compare Figure I-4).

Table I–4 Summary of maximum positions of fitted Gaussian functions in DR-UV/Vis spectra of *thin layer precursors* and the corresponding absorption edge position.

Sample	absorption maximum / cm^{-1}				absorption edge position / eV
	d-d	CT (Cu-O-Cu)	CT (O-Cu)	CT (O_{SiO2}-Cu)	
10.9Cu	13625	25506	33192	40982	3.49
14.9Cu	13505	25379	32913	40900	3.34
16.3Cu	13592	25430	33001	40999	3.32
22.3Cu	13630	25327	32544	40265	3.20
27.1Cu	13579	25266	32594	41192	3.12

Further absorption bands can be assigned to charge transfer (CT) transitions. Both the $[CuO_x]_n$ structure and chemical property of ligands influence the absorption maxima of CT transitions. The absorption maximum of ligand to metal charge transfer (LMCT) transition of O^{2-} to Cu^{2+} was located at 41000 cm^{-1}[70–72] referring to silica lattice oxygen[67] connected to a Si atom of SBA-15. Fit results of all oxidic precursors (Table I–4 and Table I–5) revealed the maximum position of this transition in the range from 40200 cm^{-1} to 42800 cm^{-1}. Large deviations in the peak position of this CT transition were caused by spectrometer-related limitations at values higher than 40000 cm^{-1}.

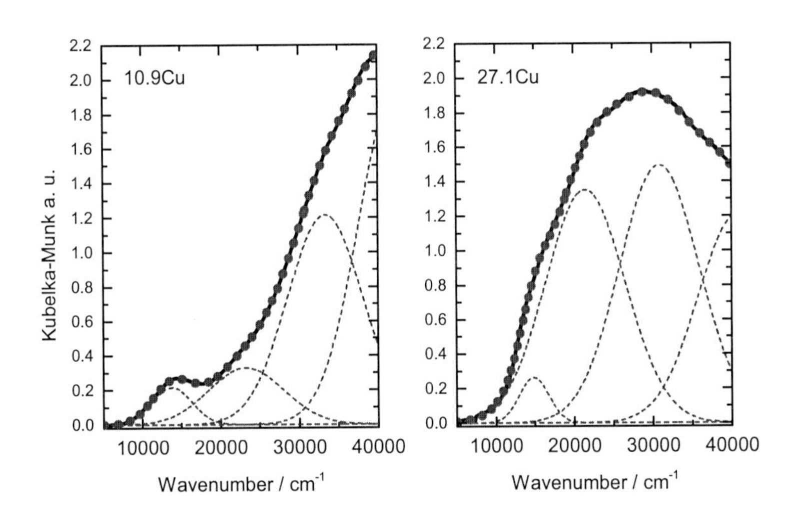

Figure I-6 DR-UV/Vis spectra (black) of 10.9Cu left and 27.1Cu right *thick layer precursors*. Blue broken lines represent Gaussian functions of single contributions of each absorption band. The fitted spectra (blue dots) overlap nearly perfectly with measured spectra (black).

The CT transition between 30000 cm^{-1} and 32000 cm^{-1} may be assigned to either square plane $[CuO_4]_n$ units condensed in chains[73] or CT from $(HO)^{-1}$ to Cu^{2+}[74]. In combination with CT transitions in the range between 22300 and 25000 cm^{-1} the absorption bands in the range between 30000 and 32000 cm^{-1} indicated planar $[Cu\text{-}(O)_2\text{-}Cu]$ dimers.[71] Hence, in *thin layer precursors*, Cu^{2+} was found in $[CuO_6]$ units and $[CuO_4]$ square plane units. Due to missing absorption bands below 10000 cm^{-1}[67], $[CuO_4]$ tetrahedron as structural motive can be excluded. As shown in paragraph *I.3.5* CuO_x particles built up by $[CuO_x]$ units tended to grow with higher Cu loadings. As result, intensities of CT absorption bands of connected $[CuO_x]$ units, *i. e.* absorption at 21500-23300 cm^{-1} and 30900-33600 cm^{-1} increased with increasing loading. Simultaneously, position of maximum absorption shifted to lower wavenumbers which can be explained by a size confinement effect.[75] Additionally, Ismagilov reported[73] a possible self-reduction of $[CuO_x]_n$ chains to structural motives like $[Cu^{2+}\text{-}O^{2-}\text{-}Cu^{+}]$, absorbing in the same range. Cu^{+} centers coordinated by O^{2-} probably existed in 22.3Cu and 27.1Cu *thick layer precursors* as indicated by XRD and XAS (compare Figure I-4 and Figure I-8). However, DR-UV/Vis spectra of references Cu_2O and CuO and spectra of oxidic precursors differed distinctly. Hence, the presence of larger crystalline copper oxide particles was unlikely, because the crystalline reference compounds Cu_2O and CuO showed a sharp absorption edge at 20200 cm^{-1} and at 13000 cm^{-1}, respectively (Appendix, Figure 0-1).

Table I–5 Summary of maximum positions of fitted Gaussian functions in DR-UV/Vis spectra of *thick layer precursors* and the corresponding absorption edge position.

sample	absorption maximum / cm^{-1}				absorption edge position / eV
	d-d	CT (O-Cu-O)	CT (O-Cu)	CT (O_{SiO2}-Cu)	
10.9Cu	13774	23292	33580	42870	3.39
14.9Cu	14319	22482	31941	41230	2.84
16.3Cu	14223	22518	32075	41327	2.92
22.3Cu	14667	21446	30858	40552	1.95
27.1Cu	14878	21514	30957	40957	1.94

I.3.7 Estimation of CuO_x particle sizes from optical band gap

The position of the absorption edge in a DR-UV/Vis spectrum might be a useful indicator for the metal oxide particle size. Shape and size of CuO particles influence the optical band gap.[76] The optical band gap of CuO at 25 °C is given as 1.35 eV.[77] According to Weber[78], the position of the absorption edge of CT bands of molybdenum oxides depends on the structural motifs. He demonstrated that an increasing linkage of $[MoO_6]$ units evoked a redshift of the absorption edge. Accordingly, in Figure I-7 the absorption edge positions of *thin layer precursors* (green triangles) and *thick layer precursors* (black squares) are summarized depending on Cu loading. Assigning the blue shift of the absorption edge compared to reference CuO to diminished particle sizes, absorption edge energies of oxidic precursors indicated the presence of nanoscale CuO particles. The decrease of the absorption edge position of *thin layer precursors* indicated a slight increase of *n* in connected $[CuO_6]_n$ units. Hence, higher Cu loadings in *thin layer precursors* resulted in slightly increased CuO_x particle size consisting of $[CuO_6]_n$ oligomers present on SBA-15. As a special exception the 10.9Cu *thin layer precursor* (3.5 eV) showed nearly the same absorption edge as the 10.9Cu *thick layer precursor* (3.4 eV). Hence, structure and dispersion of $[CuO_6]_n$ units were similar in both oxidic precursors. Otherwise, *thick layer precursors* with higher Cu contents revealed a distinct redshift (2.9 eV-1.9 eV). Chen *et al.*[79] found for CuO nanorods (4-6 nm x 50-400 nm) an optical band gap of 2.3 eV. They explained the blue shift of DR-UV/Vis absorption edge energy with a quantum confinement effect.[75]

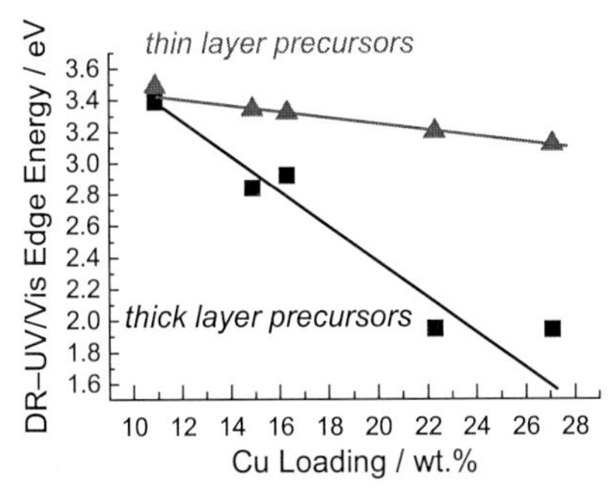

Figure I-7 Depiction of DR-UV/Vis absorption edge energies in dependency on Cu content. *Thin layer precursors* are presented in green triangles and linear fit function. *Thick layer precursors* are presented in black squares with linear fit function.

The interplay of various CuO particle shapes and various CuO particle sizes may be responsible for rough linearity found between DR-UV/Vis absorption edge position and Cu loading of *thick layer precursors* (Figure I-7, black squares). The *thick layer precursors* clearly exhibited enlarged CuO_x particles with increasing Cu loading. Interestingly, the difference of CuO_x particle sizes of *thin* and *thick layer precursors* obtained from different calcination routes increased with increasing Cu loading. *Thick layer precursors* with high Cu contents yielded CuO_x particles resembling to crystalline CuO.

I.3.8 X-ray absorption near edge spectroscopy (XANES)

Normalized Cu K edge XANES spectra of oxidic precursors and crystalline reference CuO are summarized Figure I-8. XANES spectra of *thin layer precursors* (left) were similar to each other but different from that of CuO. The absorption edge position is similar to that of Cu^{2+} in bulk CuO. However, deviations from reference crystalline CuO can be seen at 8.983 keV (Figure I-8, left). In comparison, XANES spectra of *thick layer precursors* varied with varying Cu loading. The XANES profile of 10.9Cu *thick layer precursor* resembled that of 10.9Cu *thin layer precursor*. The deviation in XANES profiles of *thick layer precursor* at 8.983 keV correlated with Cu content. This can be seen in Figure I-8, right. With increasing Cu loading, the XANES profiles of *thick layer precursors* became more similar to that of reference CuO. Moreover, the XANES spectrum of 27.1Cu *thick layer precursor* exhibited an additional slight incline at 8.980 keV (Figure I-8, top right arrow) indicating the presence of Cu^{+} (compare Figure I-4, XRD).

The first derivative of each XANES spectrum is shown in the bottom of Figure I-8, to reveal shifts of inflection points in absorption edge. The inflection points represent the maxima in the first derivatives and are listed in Table I–6. Three transitions can be assigned to the dipole forbidden but quadrupole allowed $1s \rightarrow 3d$ transition at 8.977 keV[80–82], the $1s \rightarrow 4p$ transition at 8.991 keV, and a further $1s \rightarrow 4p_z$ "shake-down" transition between 8.983 keV and 8985 keV[81,82]. In all oxidic precursors, the $1s \rightarrow 3d$ transition was found (see inset Figure I-8) and is thus decisive for the presence of Cu^{2+}.

The $1s \rightarrow 4p$ transition was found in an energy range from 8.9917 keV to 8.9907 keV for all oxidic precursors, and the CuO reference. The small variance in energies for *thin layer precursors* around 8.9916 ± 0.0001 keV corroborated the structural similarity of $[CuO_6]_n$ to each other on SBA-15 (compare DR-UV/Vis). However, $[CuO_6]_n$ units differed from those in CuO reference. Conversely, the *thick layer precursors* exhibited a shift of this transition from 8.9913 keV to 8.9907 keV with increasing Cu loading. This indicates a transformation from $[CuO_6]_n$ to a CuO like structure with increasing Cu content. In addition, this variation of $[CuO_6]_n$ structure was underlined by the more marked shift of the $1s \rightarrow 4p_z$ shakedown transition. The energy position of the $1s \rightarrow 4p_z$ transition was identical in *thin layer precursors*, while the $1s \rightarrow 4p_z$ of *thick layer precursors* was shifted from 8.9851 keV to 8.9839 keV with increasing Cu loading. Therefore, with increasing Cu content the energy position of the $1s \rightarrow 4p_z$ transition in *thick layer precursors* approached that of reference CuO, as already indicated by DR-UV/Vis.

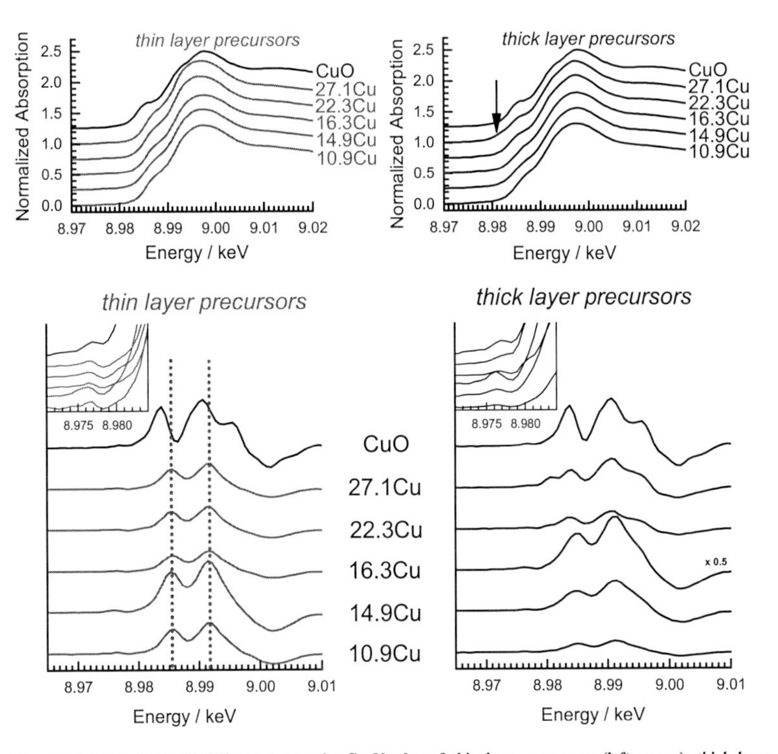

Figure I-8 Normalized XANES spectra at the Cu K edge of *thin layer precursors* (left, green), *thick layer precursors* (right, black) and CuO above, respectively, on the top. The corresponding first derivatives of XANES spectra at the Cu K edge of *thin layer precursors* (left, green) and *thick layer precursors* (right, black) in comparison to CuO reference are shown below. Numbers before *Cu* represent Cu loading. Insets show enlarged range from 8.971 to 8.984 keV. The sequence of curves is the same as in large diagrams.

The shift of $1s \rightarrow 4p_z$ transition depends on the distance between Cu and O atoms in *z* direction[83] and thus on the degree of tetragonal distortion of [CuO_6] units[81]. According to Grandjean *et al.*[84], comparison of reference CuO and *thin layer precursors* revealed a shift of about 2 eV indicating the presence of nearly square plane coordinated Cu^{2+} in *thin layer precursors*.

Table I–6 List of observed energies of electronic transitions, *i. e.* inflection points at the Cu K edge. Found edge energies of *thin layer precursors* are printed in green and on the left side in each column. Edge positions of *thick layer precursors* are printed in black and on the right side in each column. Edge energies of crystalline CuO and Cu_2O as reference are given in the last two lines on the bottom.

sample	position of maximum incline / keV		
	1s→3d	1s→$4p_z$ "shakedown"	1s→4p
10.9Cu	8.9765 / 8.9763	8.9856 / 8.9851	8.9917 / 8.9913
14.9Cu	8.9762 / 8.9764	8.9854 / 8.9848	8.9915 / 8.9911
16.3Cu	8.9765 / 8.9762	8.9856 / 8.9849	8.9917 / 8.9912
22.3Cu	8.9765 / 8.9766	8.9854 / 8.9838	8.9915 / 8.9907
27.1Cu	8.9765 / 8.9766	8.9855 / 8.9839	8.9916 / 8.9907
CuO	8.9977	8.9837	8.9905
Cu_2O			8.9803

I.3.9 Extended X-ray absorption fine structure (EXAFS)

EXAFS data analysis at the Cu K edge provides information about the neighboring atoms around the absorbing Cu atom. The $FT(\chi(k)\cdot k^3)$ of all oxidic precursors and crystalline CuO are summarized in Figure I-9. The green dotted curves represent spectra of *thin layer precursors*, and the black curves represent spectra of *thick layer precursors*.

The $FT(\chi(k)\cdot k^3)$ of *thin layer precursors* were similar to each other independent of Cu loading. Comparison of $FT(\chi(k)\cdot k^3)$ of *thin layer precursors* and *thick layer precursors* with the same Cu content revealed distinct differences in amplitude and position of peaks. The first peak at 1.5 Å of the pseudo radial distribution function was largely independent of the calcination mode and Cu content. *Thick layer precursors* showed increased amplitudes around 2.6 Å and 5.5 Å. Contributions of more distant backscattering atoms indicated crystalline CuO_x structures. Although, the difference in amplitude of the pseudo radial distribution functions at 2.6 Å or 5.5 Å seemed to be small, they are in good agreement with results from XANES and DR-UV/Vis analysis. For example, the Cu loading was less in 14.9Cu *thick layer precursor* than in 16.3Cu *thick layer precursor*. CuO_x particles were larger in 14.9Cu *thick layer precursor* indicated by absorption edge position. Higher amplitudes in pseudo radial distribution function of 14.9Cu *thick layer precursor* were observed indicating larger CuO_x particles, too. The CuO like nanocrystals were formed preferable in *thick layer precursors* exhibiting high Cu contents.

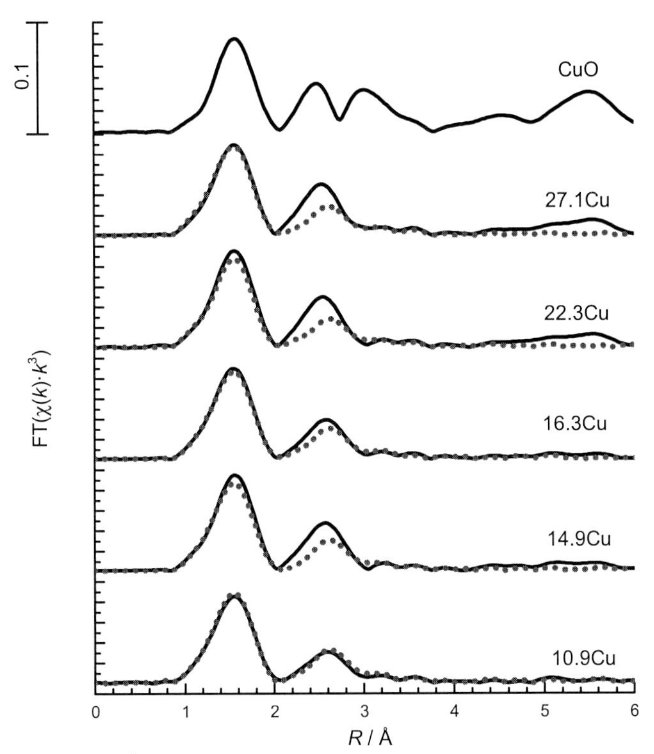

Figure I-9 FT($\chi(k)\cdot k^3$) at Cu K edge of all CuO_x/SBA-15 and crystalline CuO on the top. Number before *Cu* gives Cu loading. Black lines represent spectra of *thick layer precursors*, and green dots represent spectra of *thin layer precursors*.

I.3.10 Refinement of EXAFS spectra of thin layer precursors

EXAFS spectra were fitted with a suitable fit procedure to resolve apparent differences in $[CuO_6]_n$ structures. Starting from the local structural motif of crystalline CuO as model the numbers of parameters was reduced until measured and fitted magnitude and imaginary phase superimposed. The Debye-Waller-Factor (DWF) of all six Cu-O backscattering path of $[CuO_6]$ unit was the same. For *thin layer precursors*, four single scattering paths were necessary to obtain satisfying fit results (Figure 0-5 and Table 0–1 in *Appendix*). Starting point of the refinement was the distorted $[CuO_6]$ octahedron of crystalline CuO (Figure I-10). A tetragonal square plane was assumed with four oxygen atoms around the Cu^{2+} at two slightly deviating Cu-O distances as in CuO. Additionally, two axial Cu-O distances were introduced (dotted bonds in Figure I-10). However, the coordination number (*CN*) of the backscattering path of the axial oxygen atom had to be reduced from two to one to achieve the same DWF for all Cu-O backscattering paths. This indicates a high degree of static disorder. In addition, one Cu-Cu backscattering path with a *CN* of two was included. Cu-Si backscattering paths coming from support Si atoms were not required[54] possibly because of the higher Cu loadings in oxidic precursors (10.9 to 27.1 wt.%) compared to reference (5 wt.%)[54].

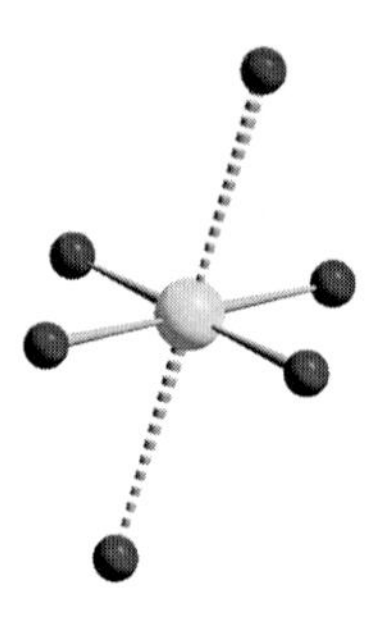

Figure I-10 Distorted [CuO_6] octahedron with Cu as central atom (turquoise) coordinated by six O atoms (red). Broken bonds represent axial Cu-O bonds. Other O atoms create a slightly distorted square plane with Cu as center atom adapted form reference[85].

The fit results of *thin layer precursors* are summarized and compared to fit results of crystalline CuO in Figure I-11 (compare Table 0–1). FT($\chi(k)\cdot k^3$) and corresponding fitted functions for 27.1Cu *thin layer precursors* are given in Figure I-12 left. Further illustrations of experimental and theoretical FT($\chi(k)\cdot k^3$) of other oxidic precursors are depicted in Figure 0-5 (*Appendix*).

Apparently, fitted parameters for *thin layer precursors* were similar to each other but showed distinct deviations from those of the CuO reference. Distances between Cu atoms and in square plane O atoms were slightly shorter (Figure I-11, top left) while distances of axial O atoms to absorber were in the same range as in reference CuO (Figure I-11, top left). The reduced *CN* and the increased DWF (Figure I-11, bottom left) indicated an increased static disorder. Conspicuously, the neighboring Cu atoms were found at distinctly larger distances (Figure I-11, top right). Furthermore, the *CN* was found to be two in oxidic precursors instead of four in reference CuO. Again, the disorder parameter DWF was increased (Figure I-11, bottom right) corresponding to varying Cu-Cu distances. Therefore, a triangular $[CuO_6]_3$ or a closed ring of $[CuO_6]_n$ chains might be present on SBA-15. Moreover, distorted octahedron could be connected differently such as face sharing, edge sharing or corner sharing. Various types of connected $[CuO_6]_x$ were accessible for the same x resulting in various Cu-Cu distances. In conclusion, the EXAFS refinement suggested various $[CuO_6]_{3\text{-}z}$ units built up from distorted [CuO_6] octahedron present on SBA-15 as majority phase.

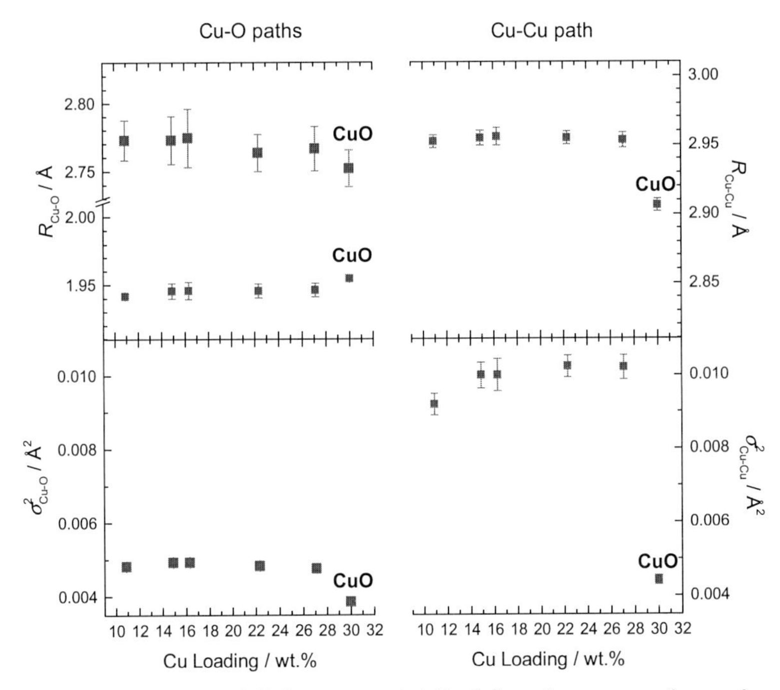

Figure I-11 EXAFS fit results of *thin layer precursors* including CuO as reference, representing error bars base on 95 % certainty.

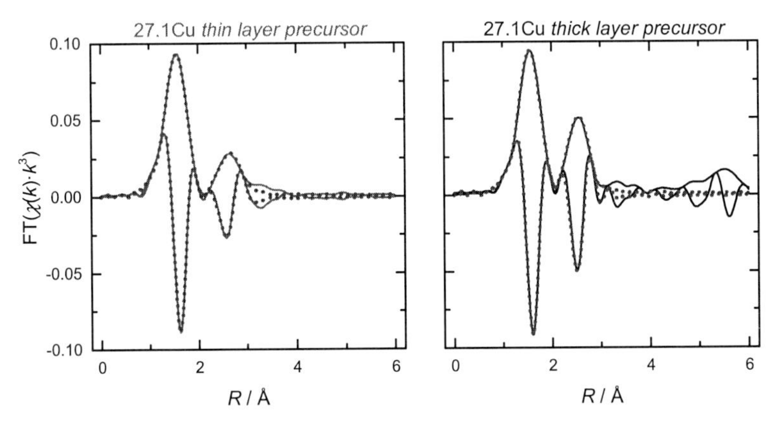

Figure I-12 Measured FT($\chi(k)\cdot k^3$) and corresponding imaginary part of 27.1Cu *thin layer precursor* and 27.1Cu *thick layer precursor* on the left in green and on the right in black, respectively. Theoretical FT($\chi(k)\cdot k^3$), and the corresponding imaginary part are shown as blue dots.

I.3.11 Refinement of EXAFS spectra of thick layer precursors

Refinement of Cu K edge FT($\chi(k)\cdot k^3$) of *thick layer precursors* revealed varying local structures of CuO_x particles depending on Cu loading. XANES analysis of *thick layer precursors* revealed larger CuO_x particles with shorter axial Cu-O distances (see above) and thus different distorted [CuO_6] units. These [CuO_6] units in CuO_x particles of *thick layer precursors* resembled that of crystalline CuO. Therefore, a structural model was assumed closed to that of *thin layer precursors*. An additional Cu-Cu distance had to be considered and increased coordination numbers (*CN*s) were necessary to model the larger CuO_x particles. The assumed *CN*s of Cu-Cu scattering paths were 2 and 0.5 for 10.9Cu, 14.9Cu and 16.3Cu *thick layer precursors*. The *CN*s of Cu-Cu scattering paths were increased to 3 and 1.5 for 22.3Cu and 27.1Cu *thick layer precursors*. However, the *CN*s of corresponding scattering paths of reference CuO were 4 for both Cu-Cu scattering paths. The good agreement of measured and calculated spectra can be seen in Figure I-12 right for 27.1Cu *thick layer precursor*. Further experimental and theoretical FT($\chi(k)\cdot k^3$) of *thick layer precursors* are depicted in Figure 0-5. Corresponding fit results are summarized in Table 0–2 and depicted in Figure I-13. Similar to *thin layer precursors*, *thick layer precursors* showed an apparent contraction of the [CuO_4] square plane (Figure I-13, top left). Distance to axial oxygen atom seemed to be increased compared to those of reference CuO (Figure I-13, middle left) and *thin layer precursors* (Figure I-13, top left). The DWF of Cu-O scattering paths was slightly increased, compared to that of reference CuO, which indicated differently distorted [CuO_6] units in *thick layer precursors* (Figure I-13, bottom left).

The reduced *CN*s of Cu-Cu scattering paths (Figure I-13, right) indicated CuO_x nanoparticles present on SBA-15. Moreover, the refined Cu-Cu distances were enlarged (Figure I-13, top and middle right). The increased DWF (Figure I-13, bottom, right) revealed higher static disorder in CuO_x particles than in CuO reference. However, the static disorder of CuO_x particles in *thick layer precursors* was smaller than that of CuO_x particles in *thin layer precursors*. Hence, the static disorder was apparently related to CuO_x particle size and with that it was related to degree of connectivity of [CuO_6] units.

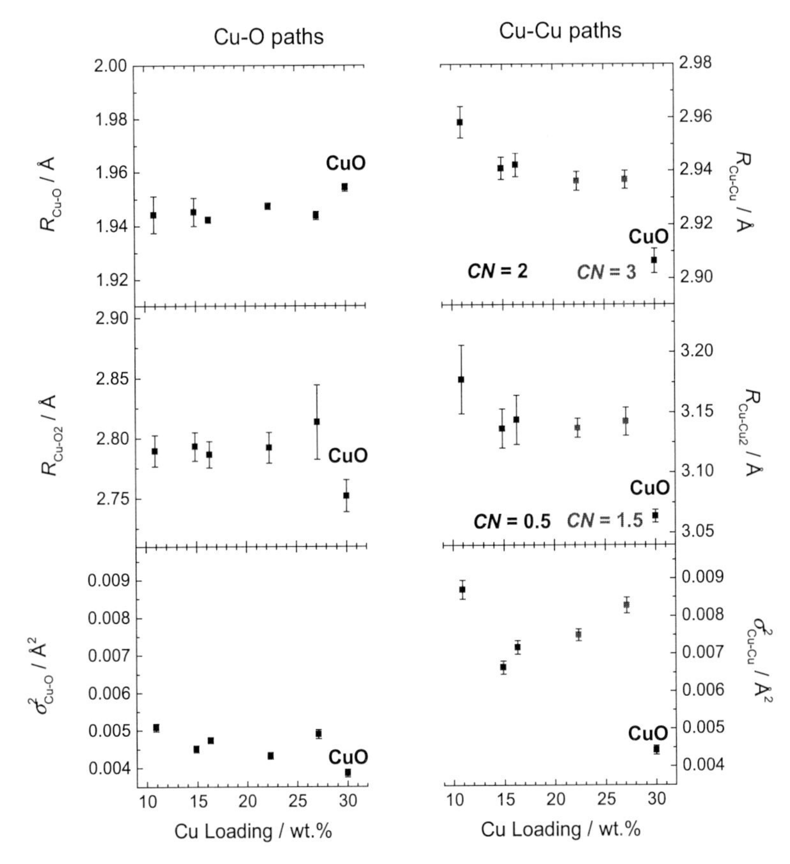

Figure I-13 EXAFS fit results of *thick layer precursors* and reference CuO. Error bars are given with 95 % certainty. σ^2 refers to all corresponding backscattering atoms independent of length of backscattering path.

I.3.12 Processes during calcination

During calcination of SBA-15 treated with Cu citrate dissolved in ammonia the citrate precursor completely decomposed and at 250 °C was transformed into CuO_x particles. Gaseous products of decomposition such as water, CO, CO_2, nitrate and probably other organic compounds were formed during calcination. Removal of these products has been shown to be a crucial step, affecting the dispersion of CuO_x particles on silica support.[26,50] For here investigated CuO_x/SBA-15 samples the following steps during removal were likely. Gaseous products had to leave the pores of a SBA-15 particle. The removal of the gaseous products is governed by diffusion process. Well spread powders of thin layers offers a great surface and gaseous products can easily be taken in atmosphere. Conversely, thick powder layers include various hollow spaces between the SBA-15 particles. Having left the pores of a SBA-15 particle, the gaseous products enrich in the hollow spaces. They are hindered to move away due to other SBA-15 particles blocking the way. Hence, a diffusion barrier is created and gaseous products cannot leave easily the pores leading to longer residence times of gaseous products. This enables interaction between Cu ions and gaseous products.

Especially H_2O was adsorbed on SBA-15 and might be thus retained in the pores of SBA-15. Complete dehydration occurred at elevated temperatures around 270 °C[86], while CuO_x/SBA-15 were calcined at 250 °C. Consequently, hydrothermal conditions were created. Under hydrothermal conditions copper ions are highly mobile, tend to migrate, and may lead to CuO_x particle growth. Furthermore, the formation of copper complexes with other gaseous products was likely. As consequence copper ion mobility increased and lead to CuO_x particle growth.[50]

An additional process during treating SBA-15 with Cu citrate dissolved in ammonia is etching of silica by the ammonia solvent. Ammonia catalyzes hydrolysis and condensation of silanol groups.[48] Therefore, the number and the distance between silanol groups of silica surface were modified during incipient wetness and subsequent calcination. With it the electrostatic attraction of copper ions by silanol groups might be changed. Moreover, formation of copper hydroxide or copper silicate compounds (*e. g.* dioptase) was possible.

I.3.13 Constitution of CuO_x particles

Depending on the calcination mode, various CuO_x particles were found on SBA-15 support. Copper ions are prone to build up strongly distorted [CuO_6] octahedron.[68] Itho and coworkers[72] presented DR-UV/Vis spectra of CuO/SiO_2 samples, which were similar to those of *thin layer precursors*. They concluded that only one Cu species was present on the silica support as octahedral [CuO_6] unit. It was also reported, that Cu atoms may occupy interstitial Si atom positions.[87] However, missing of corresponding absorption bands in DR-UV/Vis spectra excluded tetrahedral coordination of Cu^{2+} in oxidic precursors. In nature distorted [CuO_6] octahedrons can be found in several copper minerals (Figure I-14). The distorted [CuO_6] octahedron consists of a square plane including copper ion in the center (Figure I-14), and of two axial oxygen atoms at larger distances to the copper ion (Figure I-10, excluded in Figure I-14). These distorted octahedrons can be linked in various ways. For example, chains of octahedrons are found in copper silicate (Figure I-14, right)[88], corner sharing dimers were found in dioptase (Figure I-14, middle)[89], and completely edge sharing linkage is present in CuO (Figure I-14, left)[85].

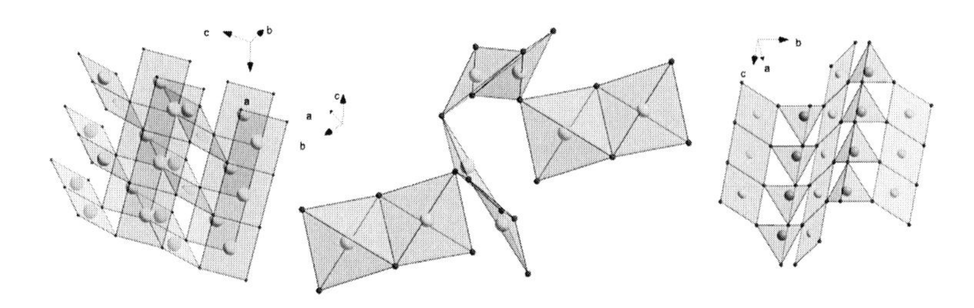

Figure I-14 Parts of the structures of CuO (left), dioptase (middle) and copper silicate (right). [CuO_4] square planes (blue) and [SiO_4] tetrahedrons in copper silicate (khaki) are shown. Other parts of structures were excluded. Distorted octahedrons arise under consideration of additionally axial oxygen atoms on the top and below of each square plane. Copper atoms are printed in blue, oxygen atoms in red, and silicon atoms in grey.

The [CuO_6] structures found on SBA-15 were derived from the [CuO_6] units of CuO as suggested by XRD, XAS, and DR-UV/Vis. EXAFS analysis yielded Cu-O distances in oxidic precursors that corresponded to those of CuO (compare Table I–7). Other minerals containing Cu^{2+} exhibit other Cu-O distances. Moreover, shortened distances were found between copper and hydroxyl ligands in comparison to Cu-O distances (compare numbers indicated with * in Table I–7). Hence, presence of copper silicates or copper hydroxide formed during synthesis can be excluded.

While the [CuO_6] units of all oxidic precursors resembled those of CuO, the size of CuO_x particles supported on SBA-15 depended on the calcination mode. *Thin layer precursors* contained small CuO_x particles while *thick layer precursors* contained larger CuO_x particles. Due to a quantum confinement effect[75], *thin layer precursors* were colored green and *thick layer precursors* were colored brown. Interestingly, in *thick layer precursors* the CuO_x particle size depended strongly on the available amount of Cu precursors.[63] Accordingly, *thick layer precursors* consisted of CuO like nanocrystals at various sizes while *thin layer precursors* exhibited disordered CuO_x particles at similar sizes.

Since the refinement of coordination number is not reliable assuming the static disorder (DWF) of reference CuO, the coordination number only presumed the number of connected [CuO_6] units. The high DWF revealed increased static disorder compared to that in CuO reference and might indicate too high coordination numbers. Thus, the major phase consisted of [CuO_6]$_{3-z}$ units resembling fragmented chains of [CuO_6] units present in *thin layer precursors*.

Table I–7 Interatomic distances between Cu atoms and O atoms found in minerals and oxidic precursors, which were derived from EXAFS refinements. * O atoms possessing a bond to a H atom.

Reference		Cu-O distance / Å	
		CuO_4 square plane	over CuO_4 plane
Cuprit Cu_2O[66]		2x 1.8483	-
Tenorit CuO[85]		2x 1.9472 2x 1.9477	2x 2.7663
Copper hydroxide $Cu(OH)_2$[90]		2x 1.9484* 2x 1.9720*	2.3563* / 2.9150*
Dioptase $Cu_6[Si_6O_{18}]\cdot 6H_2O$[89]		1.9431 1.9497 1.9600 1.9884	2.5239* / 2.6594*
Copper silicate as chain silicate $CuSiO_3$[88]		4x 1.9413	2x 2.9267
Malachite $Cu_2(OH)_2(CO_3)$[91]		1.8981* / 1.9145* 1.9107* / 1.9182* 1.9956 / 2.0486 2.0530 / 2.1099	2.5096 / 2.3694* 2.6394 / 2.3725*
CuO_x of *thin layer precursor*	10.9Cu	4x 1.942	2x 2.77
	14.9Cu	4x 1.945	2x 2.77
	16.3Cu	4x 1.946	2x 2.77
	22.3Cu	4x 1.945	2x 2.76
	27.1Cu	4x 1.946	2x 2.77
CuO_x of *thick layer precursors*	10.9Cu	4x 1.944	2x 2.79
	14.9Cu	4x 1.945	2x 2.79
	16.3Cu	4x 1.942	2x 2.79
	22.3Cu	4x 1.947	2x 2.79
	27.1Cu	4x 1.944	2x 2.81

I.3.14 Electronic structure of CuO_x particles

X-ray diffraction (XRD), X-ray absorption spectroscopy (XAS) and DR/UV-Vis spectroscopy were applied to elucidate the various structures of CuO_x particles. XRD indicated CuO like nanocrystals present in *thick layer precursors*, whereas no long range ordered CuO particles were detected in *thin layer precursors*. XANES and DR-UV/Vis indicated that the electronic structure of CuO_x particles is affected by the particle size. In Figure I-15 the shift of the $1s\rightarrow 4p_z$ "shakedown" transition is given as function of the optical absorption edge of oxidic precursors and crystalline CuO in the UV/Vis range. *Thin layer precursors* were found in top right corner as green triangles, and *thick layer precursors* are depicted as black squares. *Thick layer precursors* exhibited a linear relationship between DR-UV/Vis absorption edge energy and the inflection point of the $1s\rightarrow 4p_z$ shake-down transition.

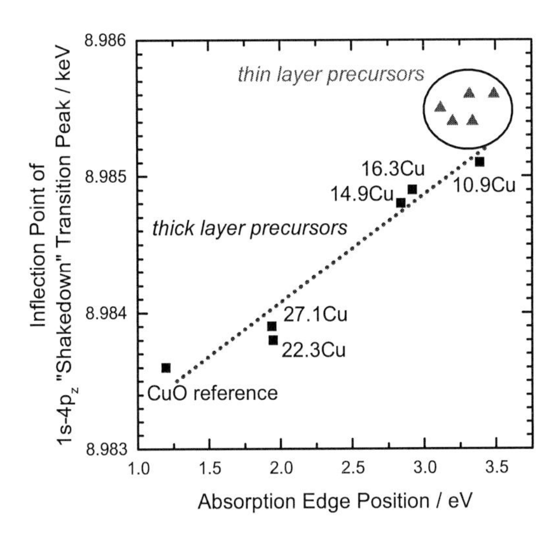

Figure I-15 Inflection point 1s-4p$_z$ "shakedown" transition of oxidic precursors against DR-UV/Vis absorption edge position of oxidic precursors and crystalline CuO reference. *Thin layer precursors* and *thick layer precursors* are depicted as green triangle and black squares, respectively. *Thin layer precursors* are circumscribed. The dotted line represents the linear fit for *thick layer precursors* including CuO reference.

Both, the surface sensitive DR-UV/Vis spectroscopy and the integral XAS revealed the same electronic structure of CuO_x particles supported on SBA-15. Hence, for future investigations of CuO_x particles supported on SBA-15, XAS and DR-UV/Vis provide the same information on CuO_x size. However, there was some uncertainty about homogeneity of CuO_x particles in *thick layer precursors*, due to overlapping signals of various structure motifs. It might be possible that CuO_x particles found in *thin layer precursors* were present in *thick layer precursors* besides larger and more ordered CuO_x particles.

I.3.15 Proposed structure of CuO_x/SBA-15 precursors

After depositing CuO_x particles on SBA-15 the typically mesoporous structure of SBA-15 was maintained. After employing the calcination without copper citrate micropores were still observed ($A_{mesopore}/A_{BET} \approx 0.92$). Even at Cu loadings of 27.1 wt.%, no mesopore blocking was observed. However, intrawall micropores in CuO_x/SBA-15 samples were not detected any more. *Thin layer precursors* exhibited less micropores ($A_{mesopore}/A_{BET} \approx 1$) than *thick layer precursors* ($A_{mesopore}/A_{BET} \approx 0.98$). Hence, it is likely that CuO_x particles in *thin layer precursors* filled up micropores or closed them. Similar behavior was observed for SiC coatings in SBA-15.[61] Open micropores in *thick layer precursors* indicated an agglomeration of [CuO_6] units, resulting in larger CuO_x particles. The mode of calcination influenced the spreading of CuO_x on SBA-15. The following schematic representation of oxidic precursors can be drawn (see Figure I-16). Cu citrate precursors were deposited in micropores of SBA-15. During calcination, CuO_x particles were formed. In *thin layer precursors* the fast removal of gaseous products of decomposition resulted in small CuO_x particles formed directly from decomposed precursors in micropores. The size of these CuO_x particles was limited by the micropore size ($\leq$ 2 nm). The increasing amount of copper led to nearly total blocking of micropores by small distorted CuO_x particles (Figure I-16, left).

Due to increased layer thickness of *thick layer precursors* in furnace, gaseous products were hindered to leave the powder bed. Hence, copper ion mobility was increased, and copper ions moved out of the micropores to form CuO_x particles in mesopores. Thus, more spherical CuO_x particles were formed in mesopores of *thick layer precursors*. (Figure I-16, right).

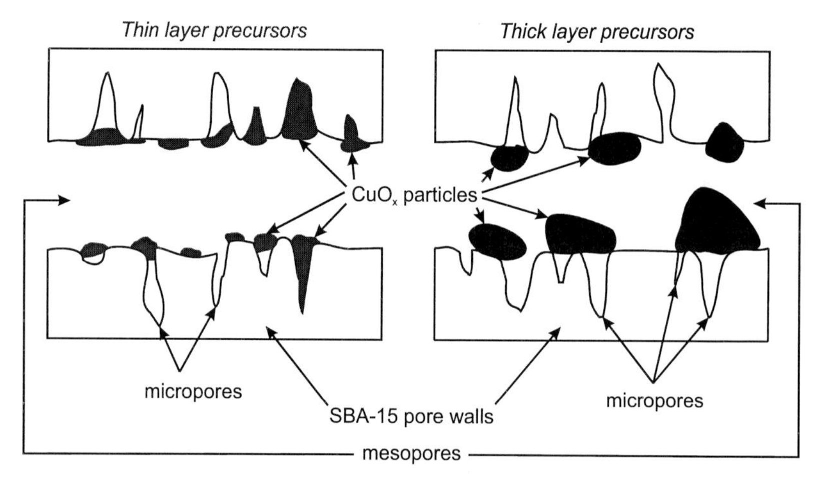

Figure I-16 Schematic structure proposal of *thin* and *thick layer precursors*, left and right, respectively. A cut along mesopores is depicted, whose walls exhibit several micropores. CuO_x particles of *thin layer* and *thick layer precursors* are shown in green and brown on the left and on the right, respectively.

Moreover, hydrolysis of Si-O bonds in pore walls might increase the amount of silanol groups and thus immobilize copper ions. Linkage of two surface silanol groups via a copper ion in the mouth of micropores would explain their pore closing and the following stabilization of pore walls. Hydrolysis of Si-O bonds is reversible and may be prolonged, if water is present in the pores. In *thin layer precursors* rapid removal of water quenched hydrolysis and preserved the R-Si-(O-Cu-O)$_n$-Si-R structure on the top of micropores. Conversely, in *thick layer precursors* the hydrolysis took more time to proceed. Hence, copper ions migrated from the micropores to growing CuO_x particles in the mesopores. Consequently, more open micropores were observed, accompanied by larger CuO_x particles in *thick layer precursor*.

I.3.16 Comparison to other supported metal oxide particles

For a given support material, the structure of supported metal oxide particles is governed by coverage of support surfaces with of MeO_x particles.[59] Low loadings of metal oxides leads to isolated [MeO_x] units. With increasing loading the [MeO_x] units are prone to form oligomers, and MeO_x nanocrystals.[59,92] These various MeO_x units showed different activities in selective propene oxidation. Furthermore, the interaction between the metal ion and the support material influences the formation of various MeO_x particles.[93] For example V oxides supported on SBA-15 showed various structures depending on hydration state of SBA-15 and on V loading. V_2O_5 nanocrystals were formed at loadings of about 7 wt.% V on SBA-15.[94] Conversely, amorphous CuO_x particles supported on SBA-15 could be obtained even for Cu contents between 10.9 and 27.1 wt.% in *thin layer precursors*. These CuO_x particles showed a similar degree of connectivity of [CuO_6] units, *i. e.* the presence of $[CuO_6]_{3-z}$ units. According to previous investigations of supported VO_x and MoO_x particles isolated [CuO_6] units may be accessible by further reducing the Cu loading to ~ 1 wt.% or less. For *thick layer precursors* CuO nanocrystals were detected at Cu contents higher than 14.9 wt.%.

Preventing the formation of crystalline metal oxides at loadings above 10 wt.% may also be desirable for other metal oxides. This can be reached by tailoring the precursor layer thickness during calcination and may be interesting for future investigations.

I.4 Conclusions

Well dispersed CuO_x particles supported on SBA-15 were synthesized using the citrate route. The incipient wetness method was employed to deposit Cu loadings of 10 wt.% to 27 wt.%. During calcination, the precursor layer thickness affected the properties of CuO_x/SBA-15 samples. Detailed structural analysis of both, mesoporous support, and deposited CuO_x particles revealed the stability of the SBA-15 support. Furthermore, the structure of the CuO_x particles was sensitivity towards calcination conditions. CuO_x/SBA-15 samples revealed distinct changes of CuO_x particle size and electronic structure as consequence of different calcination modes. Apparently, the layer thickness of powders during calcination affected the removal of gaseous products during calcination. Thicker powder layers retarded the removal of gaseous products in comparison to thin powder layers where gaseous products were released quickly. Hence, high concentration of gaseous products in *thick layer precursors* increased the mobility of Cu ions and thus supported the growth of CuO_x particles. Therefore, migration of copper ions from micropores to mesopores was deduced. CuO_x particle sizes of *thick layer precursors* increased with increasing Cu content. In contrast, the sizes of CuO_x particles of *thin layer precursors* were hardly affected by Cu loading. Moreover, CuO_x particles of *thin layer precursors* were deposited mainly in micropores. Conversely, in *thick layer precursors* the larger CuO_x particles were formed in mesopores.

Apparently, various CuO_x particles consisting of connected $[CuO_6]$ units were accessible by simple variation of calcination conditions. Copper oxides are active in selective oxidation of methanol[95] or propene[26]. Hence, CuO_x/SBA-15 is a suitable model catalyst to reveal correlations between $[CuO_x]_n$ structures and catalytic activities. Moreover CuO_x/SBA-25 samples may reveal relationships between electronic and geometrical structure. With focus on oxidation state of copper atoms and special $[CuO_x]_n$ structure motives, informative correlation might be accessible.

Furthermore, controlling the synthesis of various CuO_x particles is a good basis for synthesizing structurally different Cu metal particles with regard to methanol chemistry. These copper metal particles may be obtained by reductive activation and probably exhibit different microstructural characteristics. The ability of adjusting the size or the microstrain of Cu metal particles would open up a new field for systematic investigation. The new aspect would be the comparison of differently structured Cu metal particles supported on SBA-15 while the chemical composition is the same. Hence, the next chapter deals with structure activity correlations between Cu metal particles, obtained from *thin* and *thick layer precursors*, and H_2 formation by methanol steam reforming (MSR).

Chapter II Formation, structural and catalytic characterization of methanol steam reforming Cu/SBA-15 catalysts

II.1 Introduction

Cu containing catalysts are frequently used in methanol chemistry. In the 1980s it has been demonstrated that Cu metal particles of complex ternary catalysts play the most important role in well performing catalysts.[7,96] Furthermore, the catalytic activity correlated with the Cu surface area. However, deviations from this simple correlation were reported.[10,19] For further investigations, Cu metal particles have been deposited on several support materials such as ZrO_2[21,23], ZnO[19,57], and SiO_2[16] or even more complex like the industrially used $Cu/ZnO/Al_2O_3$. These investigations were aimed at elucidating the promoting effects or stabilizing effects of support materials on Cu metal particles. Sà summarized[5] previous examination of a wide range of differently composed catalysts for methanol steam reforming. The presence of reduced Cu metal particles in catalysts is a prerequisite for highly active catalysts. The chemical composition of support or promoter affected the dispersion[72] and partly promoted the intrinsic activity[21]. The $Cu/ZnO/Al_2O_3$ catalyst has been the basis for detailed studies on model catalysts to elucidate correlations between microstructure of Cu metal particles and activity in methanol steam reforming (MSR).

Catalytic activity depends on chemical composition[5,8,19] and on conditions in various steps during preparation such as drying, aging[45,46] or calcination[5,10,12]. The preparation route proceeds mostly via a precursor preliminary stage. Subsequent calcination to oxidic precursor is followed by activation in H_2 to form Cu metal particles.[97] It was found that good distribution of Cu and Zn atoms in CuO/ZnO oxidic precursors lead to well dispersed Cu metal particles after activation in corresponding Cu/ZnO catalysts. Hence, the distribution of Cu and Zn atoms in oxidic precursors correlated with later MSR activity of working Cu/ZnO catalysts.[18,45] These correlations are united in the term *chemical memory effect*.[18,45,98]

Especially studies on industrially relevant $Cu/ZnO/Al_2O_3$ and Cu/ZnO catalysts revealed the high complexity of catalysts for methanol chemistry.[19,45,98] Different ratios between Cu and Zn atoms in catalysts composition led to differently active Cu metal particles after activation.[19] Moreover, the interface of Cu metal particles and ZnO particles affected the microstructure of Cu metal particles.[19] The microstrain in Cu metal particles was attributed to epitaxial growth of Cu on ZnO interfaces.[11] Increased microstrain correlated with increased intrinsic activity of Cu catalysts.[19] As consequence, also defects resulting in microstrain influenced the electronic structure of Cu metal particles[99] and thus led to variation in adsorption energies on the Cu particle surface[99,100]. The electronic structure of Cu metal particles is also affected by removal of electrons, *i. e.* partial oxidation of the Cu metal particles. Oxidized Cu metal particles showed an increased MSR activity.[14,16] Combination of both electronic structure of strained Cu metal particles and shift of Fermi energy due to oxidation, may affect performance of catalysts.[101] The electronic structure of metallic nanoparticles is further governed by quantum confinement effect.[102] With decreasing number of atoms per particles, the metallic character of nanoparticles is disbanded. Moreover, smaller particles consist of a larger fraction of surface atoms. Surface atoms exhibit an incomplete coordination spheres and have thus other electronic structures than bulk atoms. The reduced particle size determines also the number of atoms forming corners, edges, kinks, or plane faces. In heterogeneous catalysis, the geometrical structure of surfaces governs adsorption energies and therefore reaction pathways, *e. g.* bond cleavage or elimination.[103]

The frequently used ZnO promotes the Cu metal particles by preventing them from sintering.[10] Additionally, ZnO induced stacking faults in Cu metal particles.[6] Furthermore, strong metal support interactions were discussed for Cu/ZnO[104] and Cu/ZnO/Al_2O_3[105]. However, ZnO may influence both, the structure of Cu metal particles and the catalytic activity by forming various active centers at the Cu/ZnO interface[6] enabling spillover of H_2[5]. Here, investigations focused on the interplay between Cu metal particle size, the microstructure of Cu metal particles, and the catalytic activity in methanol steam reforming. Elucidating the effects of the microstructure of Cu metal particles on catalytic activity required omitting ZnO in catalysts composition. The mesoporous silica, SBA-15, sufficed to prevent Cu metal particles from sintering[25,50] and particle shape transformation[106]. Silica was found to be inert in methanol steam reforming, and to stabilize Cu metal particles. Starting from the oxidic precursors introduced in *Chapter I* , active Cu metal particles were obtained by reduction of CuO_x particles of oxidic precursors. The microstructure of Cu metal particles and the performance of catalysts were investigated. Furthermore, structural characteristics of CuO_x particles in oxidic precursors were correlated to those of Cu metal particles in catalysts with regard to *chemical memory effect*.

II.2 Experimental

II.2.1 Temperature-programmed reduction and determination of Cu surface area

First, a temperature-programmed reduction (TPR) baseline of SBA-15 was measured to correct for possible contributions from SBA-15 in the following measurements. Sample powders, placed between silica wool, were deposited in a silica glass tube reactor. The reactor was connected to a BELCat-analyzer. The reaction gas flew over the sample and subsequently over a molecular sieve before H_2 consumption was measured by a TCD. Here, 4 Å molecular sieves were used for trapping H_2O formed during TPR.

The following reaction sequence was performed to estimate the Cu surface area, including TPR profiles. First, the oxidic precursors were reduced to obtain Cu metal particles on SBA-15. Therefore, 30 mg of sample were reduced in a flow of 40 ml/min 5 % H_2/Ar. After purging for 30 min, the reactor was heated to 250 °C at 5 °C/min for a dwelling time of 2 h. Cooling in a He flow of 80 ml/min preserved the highly oxygen sensitive Cu metal particles on SBA-15 support from oxidation. Subsequently, N_2O pulses were employed, to ensure short reactions times between gaseous N_2O and the Cu surface.[107] A He flow of 50 ml/min transported a defined volume of N_2O out of a loop over the reduced samples at 35 °C to avoid bulk oxidation.[108,109] A 13 Å molecular sieve in a trap was used for retarding N_2O during pulse experiments.[110] Thus, separation of remained reactant N_2O and freshly formed N_2 was achieved. N_2O pulses were repeated four times. Finally, calibration was done by injection of five pulses of pure N_2 under the same conditions. Subsequently, TPR was carried out again[108] according to parameters described above. Samples were held for a dwelling time of 15 min at 250 °C. Since overestimation of Cu surface areas due to partial reduction of ZnO was reported for Cu/ZnO catalysts[107] absence of ZnO in the here investigated catalysts led to more reliable Cu surface areas. Monti and Baiker[111] suggested reaction parameter K in the range of 55-140 s and P in the range of 0-20 K to suppress influences of mass transport or heat transport. Reaction parameters could not be considered, because the capacity in reactor was limited according to amount of catalysts. For here investigated Cu/SBA-15 catalysts the setup K was in the range of 34.6-86.0 s and P was in the range 1.9-7.2 K (compare Table 0–3). Experiments were thus performed using similar amount of catalysts.

Cu surface areas were calculated from the amount of created N_2 and from the amount of consumed H_2 during the last TPR. Cu surface areas obtained from created N_2 strongly spread while the Cu surface areas obtained from repeated TPR exhibited distinct smaller spreading and sufficient reproducibility. N_2 based Cu surface areas were smaller or similar to TPR based

Cu surface areas. Therefore, Cu surface areas calculated from released N_2 were neglected. The stoichiometric factor was two and surface atom density was taken as $1.46 \cdot 10^{19}$ atoms/m^2 reported in references.[107,112] Cu atom densities has been averaged on basis of atom density of Cu (100), (110), and (111) surfaces. Resulting space demand of a Cu atom was 0.68 nm^2 per Cu atom.

II.2.2 Methanol steam reforming (MSR)

Activity of catalysts was measured with the experimental setup shown in Figure II-1. Catalysts were diluted with BN (Alfa Aeser, 99.5 %, 325 meshes) to adjust methanol conversion below 15 %. Typically, 70 mg sample composed of BN and oxidic precursor were filled in a fixed bed silica tube reactor with 10 mm in diameter. The oxidic precursors were activated in a stream of 40 ml/min consisting of 5 % H_2 balanced by He. Temperature was raised to 250 °C at 5 °C/min and held for a dwelling time of 2 h. The activated catalysts were tested for 12 h at 250 °C in a stream of 2 % MeOH and 2 % H_2O balanced by He. Gas flow was adjusted to 80 ml/min resulting in a GHSV of 416 min^{-1}. The reactor was heated to 250 °C at 5 °C/min. Gas supply was adjusted by mass flow controllers. External thermostats ensured a constant temperature in the saturators for MeOH (12 °C) and H_2O (35 °C) during catalytic tests. Gas supply tubes were heated to 120 °C. Emitted reaction gas was monitored by quadrupole mass spectrometer (Omnistar, Pfeiffer) and reactant and product mixture was analyzed using a micro gas chromatograph (CP4900, Varian/Agilent). Separation of H_2 and CO from He was achieved over a 5Å molecular sieve capillary column. A prior installed backflush module was used to avoid contamination from other components. Reaction feed, *i. e.* H_2O and MeOH, as well as CO_2, methyl formate, and formaldehyde were separated over a CP-Sil8 capillary column. Chromatographic methods were optimized with regard to short analysis cycles. According to Madon and Boudart[113], influences on rate of catalysts evoked by mass transport or heat transport limitations were excluded (compare Table II–1). Conversion rates of MeOH were adjusted to be similar for all catalysts (diluting catalyst with BN).

Table II–1 Comparison of weighed-in mass of 22.3Cu *thin layer catalysts* powder, diluent BN, and mass of mixture in reactor with measured conversion rates of MeOH after 1 h time on stream (TOS).

mass catalysts	mass BN	mass in reactor	conversion rate of MeOH / $l \cdot h^{-1} \cdot g_{Cu}^{-1}$
5.1 mg	65.6 mg	68.1 mg	1.38
8.4 mg	63.4 mg	69.9 mg	1.37
8.3 mg	63.8 mg	69.2 mg	1.42

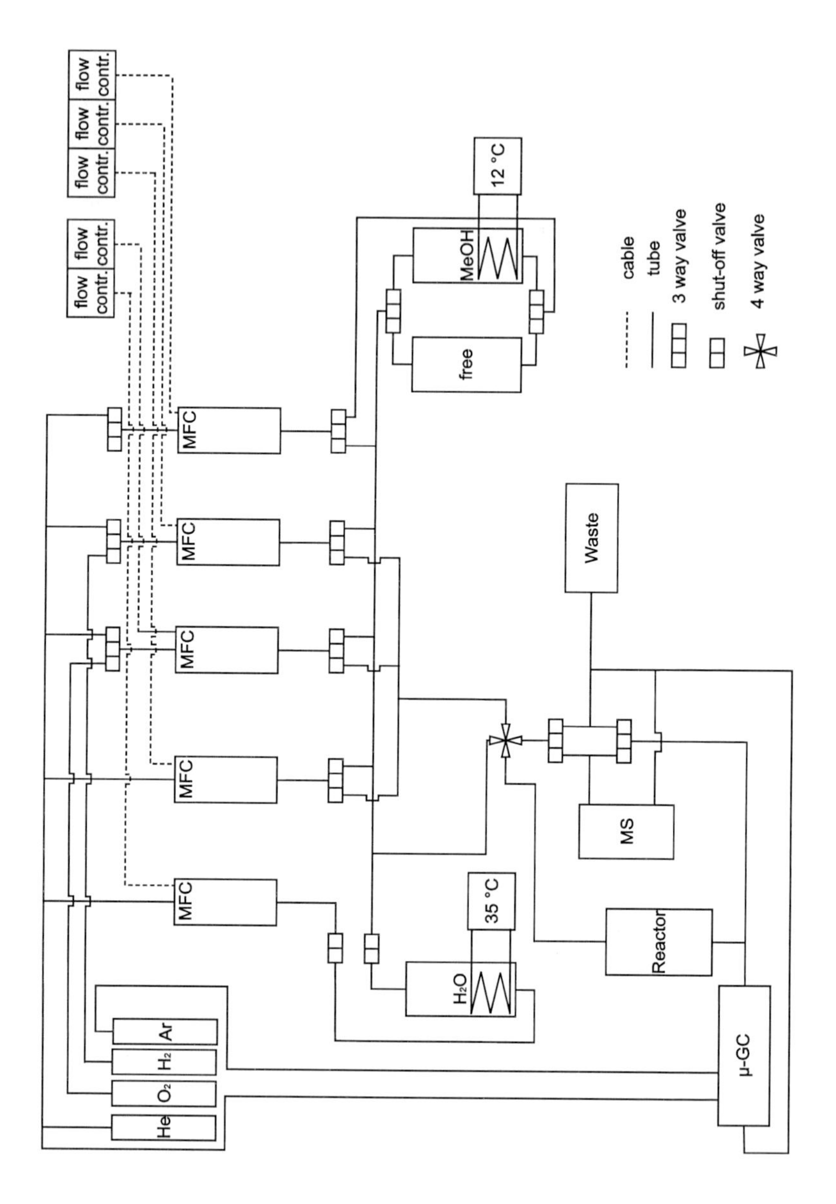

Figure II-1 Flow scheme of methanol steam reforming experimental setup.

II.2.3 In situ X-ray diffraction (XRD)

X-ray diffraction pattern of Cu/SBA-15 catalyst were measured on a STOE diffractometer during activation (TPR) and methanol steam reforming in reflection mode using Cu K_α radiation. The diffractometer used the θ/θ geometry and was equipped with a scintillation detector. Samples were deposited on silica wool in the sample holder in the center of an AntonPaar furnace. Before starting the temperature ramp, the range between 32 ° to 44 °2θ was scanned with a step size of 0.1 °2θ and a measuring time of 90 s per step at 50 °C. After heating at 5 °C/min to 250 °C in 5 % H_2 balanced by He, the range between 35 ° < 2θ < 59 ° was scanned using steps of 0.1 ° 2θ and a measuring time of 90 s per step, in total 6 h (second range). Subsequently, the furnace was cooled to 50 °C and samples were scanned in the second range. Methanol steam reforming feed was adjusted to 100 ml/min, consisting of 2 % MeOH and 2 % H_2O, balanced by He. The furnace was heated to 250 °C at 5 °C/min. At 250 °C catalysts were scanned again in the second range.

Phase analysis was done using the WinXPow software package and included data base. Reference components Cu[114], Cu_2O[66] and CuO[85] were used for phase identification. Peak shape analysis was performed with the WinXAS[60] software package. Due to data quality, many XRD fitting methods such as Rietveld refinement or diffraction peak deconvolution[24,115] led to unstable fits or inconsistent parameters and were thus not applicable for here investigated Cu/SBA-15 catalysts. Hence, the Cu(111) and Cu(200) diffraction peaks were fitted using two pseudo-Voigt-functions (pVfs) for each peak because $Cu_{K\alpha2}$ radiation could not be subtracted sufficiently. Contribution of $Cu_{K\alpha1}$ and $Cu_{K\alpha2}$ were correlated in the typical way, *i. e.* variation of peak position due to different wavelengths depending on diffraction angle and variation of peak height due to particular intensity ratio of $Cu_{K\alpha1}/Cu_{K\alpha2} = 2/1$. Finally, the height and positions of diffraction peaks, the full width at half maximum (FWHM) and the contribution of Gaussian function were refined. FWHM and part of Gaussian functions were correlated to the same value for both peaks. Appropriate background was estimated using a third degree polynomial. Start parameters for background refinement were chosen on the basis of a suitable fit of XRD pattern of SBA-15. Average crystallite sizes were calculated using Scherrer equation.[116] According to de Keijser *et al.*[117], microstrain of Cu metal particles was determined using the proposed empirical equations:

$$S = \frac{G}{4tan\theta} \qquad G = c_0 + c_{\frac{1}{2}}(1 + c\alpha)^{\frac{1}{2}} + c_1\alpha + c_2\alpha^2$$

G - Gaussian part; α - *shape determining factor*; S - microstrain; θ - diffraction angle

$c_0 = 0.184446$ $c_{1/2} = 0.812692$ $c_1 = -0.659603$ $c_2 = 0.445542$ $c = 0.998497$.

II.2.4 In situ X-ray absorption spectroscopy (XAS)

Experiments were performed in an *in situ* cell[118] at beamline X at HASYLAB at DESY, Hamburg. Therefore, sample powders and BN were mixed and pressed with a force of 0.5 t for 10 s into pellets with 5 mm in diameter and a mass of 35 mg. Sample mass was calculated to obtain an edge jump $0.9 < \Delta\mu(d) < 1.2$. XAS experiments of Cu/SBA-15 catalysts were carried out with 10.9Cu, 14.9Cu and 16.3Cu *thin* and *thick layer catalysts*. The XAS spectra were measured at the Cu K edge in the range of 8.900 keV to 9.940 keV in transmission mode using a Si(111) double crystal monochromator. QEXAFS mode was applied during heating in the range between 8.920 keV and 9.300 keV resulting in 30 s measuring time for each QEXAFS scan.

Data reduction and analysis were performed using the software package WinXAS.[60] Data analysis was performed according to reference[37]. Background subtraction and normalization were performed using a linear fit in the pre-edge range from 8.900 keV to 8.950 keV and a third degree polynomial in the post-edge range between 9.105 keV and 9.930 keV of a X-ray absorption spectrum. Smooth atomic background $\mu_0(k)$ was calculated with cubic splines in the range of $4\ \text{Å}^{-1} \leq k \leq 13\ \text{Å}^{-1}$.[119] Extracted $\chi(k)$ was k^3 weighted and then Fourier transformed into R space using a Bessel window.

II.3 Results and discussion

Samples are denoted as follows: After reduction of CuO_x particles to Cu metal particles supported on SBA-15 of *thin layer precursors* and *thick layer precursors*, samples are named *thin layer catalysts* and *thick layer catalysts*, respectively.

II.3.1 Activation of oxidic precursors during TPR up to 250 °C

Starting from the oxidic precursors, CuO_x/SBA-15, activation led to Cu nanoparticles supported on SBA-15. This transformation proceeded in 5 % H_2 atmosphere under ambient pressure at temperatures up to 250 °C. H_2 consumption of *thin layer precursors* and *thick layer precursors* during activation is given in Figure II-2. Enlarged H_2 consumption with increasing Cu loading was observed because the mass of catalysts was kept constant for all measurements. The *thin layer precursors* showed similar H_2 consumption peak profiles (Figure II-2, top). Reduction of CuO_x particles of *thin layer precursors* started at 150 °C. The highest H_2 conversion was located between 208 °C and 221 °C, and decreased to nearly zero in the range of 220 °C and 240 °C.

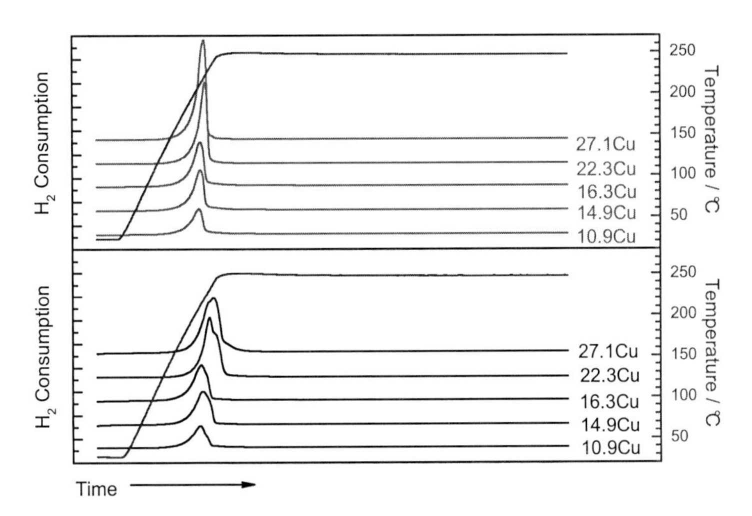

Figure II-2 H_2 consumption during activation of *thin layer precursors* (top) and *thick layer precursors* (bottom). Activation was performed in 5 % H_2 heating to 250 °C at 5 °C/min. Temperature was held for 2 h.

H_2 consumption peak profiles of *thick layer precursors* were different compared to the corresponding *thin layer precursors*. Furthermore the H_2 consumption peak profiles of *thick layer precursors* varied with varying Cu loading, but those of *thin layer precursors* were similar to each other (Figure II-2, bottom). The onset temperature of reduction was slightly increased to 155 °C. H_2 conversion increased moderately to a maximum between 212 °C and 238 °C. The subsequent decline showed a little shoulder for 10.9Cu, 14.9Cu and 16.3Cu. The shoulder developed into a second overlapping peak in TPR profiles of 22.3Cu and 27.1Cu. H_2 conversion descended to zero in the range of 230 °C to 250 °C except for 27.1Cu. The CuO_x particles of 27.1Cu *thick layer precursor* were completely reduced to Cu metal particles during dwelling time at 250 °C (compare XAS in *II.3.5*, XRD in *II.3.6* below).

TPR profiles of CuO/SiO_2 shown in the literature exhibited similar shapes of H_2 consumption peaks.[17,120] TPR profiles of supported CuO particles often show more than one peak. A common explanation for two peaks in TPR curves is the bimodal size distribution of CuO particles.[10] A relationship between temperature of maximum in H_2 consumption and size of formed Cu metal particles is published.[121] For example, CuO particles of unknown size supported on SiO_2 showed a maximum in H_2 consumption at 236 °C. The resulting Cu metal particles supported on SiO_2 exhibited Cu particle sizes between 3-4 nm. A maximum of H_2 consumption at 262 °C resulted in Cu metal particles exhibiting particle sizes of 10 nm. Therefore, the size of Cu nanoparticles of *thin layer catalysts* was estimated to be smaller than 3 nm. Accordingly, *thick layer catalysts* consisted of two types of Cu metal particles varying in size. The size of one type of particle was located between 2 and 5 nm and the size of the other type of particles was in the range of 5 nm to approximately 8 nm. Especially the 22.3Cu and 27.1Cu *thick layer catalysts* exhibited larger Cu metal particles (Table II–2).

Table II–2 Maxima of H_2 consumption peak of oxidic precursors and the estimated particle size of formed Cu metal particles according to reference[121].

sample	*thin layer catalysts*		*thick layer catalysts*	
	T_{max} / °C	d / nm	T_{max} / °C	d / nm
10.9Cu	208	< 3	212	3-5
14.9Cu	210	< 3	215	3-5
16.3Cu	210	< 3	212	3-5
22.3Cu	221	< 3	230	2-5 and 6
27.1Cu	218	< 3	238	2-5 and 8

Beside the particle size, also the increasing order of crystalline CuO particles shifted the maximum of H_2 consumption during TPR to higher temperatures.[122] This effect competed with that of the particle size and thus disentangling both was complicated. Hence, the here observed front tailing of TPR peaks may be understood as indication of inhomogeneity in particle size distribution or by differently ordered reducible CuO_x species present on SBA-15. Unfortunately, these particle sizes were estimations based on single experiments and therefore a comparison with results of complementary methods such as XRD or XAS was still needed[123] to elucidate the structural characteristics of the various Cu/SBA-15 catalysts.

Furthermore, different particle sizes of metal oxides may exhibit different reduction rates, due to the different ratio between surface atoms and bulk atoms. Consequently, the two step reduction of CuO proceeding via intermediate Cu_2O to Cu[97] (compare *II.3.2*) could not be resolved by TPR in *thin layer precursors*. During reduction of larger CuO_x particles of *thick layer precursors* (compare *I.3.5* and *I.3.13*) additional diffusion processes may become rate determining[124] and the reaction was slower.

II.3.2 XAS at Cu K edge of oxidic precursors during TPR and application of linear combination of reference XANES spectra

The XANES range of XAS spectra provides information on oxidation state of the absorber and its nearest neighbors, *i. e.* type of neighboring atoms, coordination sphere, and in special cases adsorbed molecules.[125] X-ray absorption at the Cu K edge is additionally sensitive towards Cu particle size[35] and particle shape[36]. It is a common method to determine the evolution of the phase contributions during reaction of solids from XAS measurements.[35] Here, the evolution of the Cu K edge during TPR was measured employing QEXAFS scans. Calibration, background subtraction, and normalization of each spectrum were performed prior to analysis. Principal component analysis[60] of normalized spectra, suggested three necessary phases for suitable simulation of measured spectra. Therefore, spectra of CuO, Cu_2O, and Cu metal foil were chosen as references. Using linear combination (LC) of reference spectra was already introduced in reference[97]. XANES spectra were fitted in the range of 8.950 keV to 9.000 keV and the sum of the contribution of references was constrained to be 1. Results for *thin layer catalysts* are depicted in Figure II-3. Selected parameters are listed in Table II–3. XANES spectra of CuO_x particles on SBA-15 resembled those of crystalline CuO (compare *I.3.8*). Reaching 170 °C, reduction to Cu_2O as intermediate was observed. At about 200 °C a maximum of Cu_2O content was reached and subsequent reduction to Cu metal particles started (Table II–3). During dwelling time at 250 °C, the phase composition was invariant, although the reduction was apparently incomplete. Apparent traces of CuO were mathematical artefacts due to inadequate reference spectra, and can therefore be neglected. However, apparent contributions of Cu_2O were found after reduction. This might be caused by small Cu metal particles present on SBA-15, exhibiting a size of about 2 nm (compare following paragraphs). Hence, the uncertainty for this procedure was increased due to inadequate reference spectra of crystalline phases compared to the Cu nanoparticles present on SBA-15.

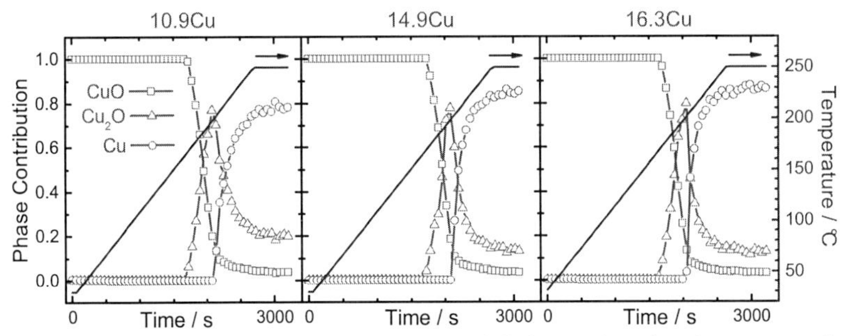

Figure II-3 Phase contributions derived from LC-XANES fits of *thin layer catalysts* using reference spectra of Cu (circles), Cu_2O (triangles), and CuO (squares) during TPR at 5 °C/min to 250 °C in 5 % H_2 balanced by He.

In comparison to CuO_x particles of *thin layer precursors*, CuO_x particles of *thick layer precursors* showed similar behavior during TPR (Figure II-4 and Table II–3). Initially present CuO_x particles were reduced to intermediate Cu_2O at 170 °C. Again further reduction to Cu started at 200 °C. Similar to *thin layer catalysts*, reduction ended within dwelling time at 250 °C. Remnants of CuO were mathematical artefacts. The contribution of Cu_2O after activation was reduced compared to *thin layer catalysts*. This indicated that reference spectra became more suitable. Cu metal particles on SBA-15 of *thick layer catalysts* were larger than those of *thin layer catalysts* because XANES spectra of Cu particles of *thick layer catalysts* resembled more the XANES spectra of Cu foil than XANES spectra of *thin layer catalysts* (Figure II-6).

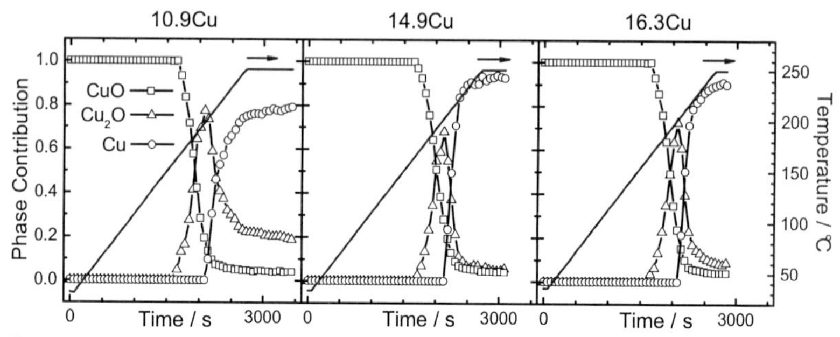

Figure II-4 Phase contributions derived from LC-XANES fits of *thick layer catalysts* using reference spectra of Cu (circles), Cu_2O (triangles), and CuO (squares) during TPR at 5 °C/min to 250 °C in 5 % H_2 balanced by He.

CuO_x particles of oxidic precursors were transformed via a Cu_2O like intermediates to Cu. A similar behavior has been reported for Cu metal particles of Cu/ZnO catalysts.[11] In contrast, direct transformation of bulk CuO and CuO films to metallic Cu under ambient atmospheres was reported by Rodriguez *et al.*[126] during isothermal reduction. They proposed that CuO being rich in defects simplified reduction resulting therefore in fast formation of Cu nuclei. The Cu nuclei activated dissociatively H_2, which led to fast formation of metallic Cu, while intermediate Cu_2O was not observed.[126] The low reduction temperatures of oxidic precursors suggested thus that the CuO_x particles supported on SBA-15 were small and disordered. The disorder may be size related.

Table II–3 Onset temperatures of CuO_x/SBA-15 samples during activation. Temperatures of maximum Cu_2O fraction is shown in line 2. Phase fractions are listed at certain temperatures given as index.

	10.9Cu		14.9Cu		16.3Cu	
	thin layer	*thick layer*	*thin layer*	*thick layer*	*thin layer*	*thick layer*
T_{onset} / °C	168	162	169	169	172	173
$T_{maxCu2O}$ / °C	197	198	198	203	201	203
Cu_2O_{max}	0.77	0.77	0.78	0.68	0.80	0.73
Cu_2O_{250}	0.2	0.19	0.13	0.06	0.13	0.09
CuO_{250}	0.04	0.04	0.04	0.04	0.03	0.04
Cu_{250}	0.78	0.79	0.86	0.93	0.87	0.91

II.3.3 Oxidation state of Cu metal particles supported on SBA-15 during TPR and MSR

Because increased methanol steam reforming activity was correlated to partly oxidized Cu metal particles[14,16], it is of interest to determine the average oxidation state of Cu metal particles during methanol steam reforming (MSR). Although the Cu particle size dominantly affected also the XANES profiles (see below in *II.3.4*) the average oxidation state was determined for here investigated Cu/SBA-15 catalysts. According to method described in *X-ray absorption near edge spectroscopy (XANES)*, analysis of Cu K edge XANES spectra of Cu/SBA-15 catalysts revealed average oxidation states of Cu metal particles between 0.1 and 1 (Figure II-5). The reduction continued during activation in H_2 and at the beginning of MSR indicating little changes in Cu metal particles. Interestingly, the deviation of average oxidation state of Cu metal particles compared to Cu foil was smaller in *thick layer catalysts* than in *thin layer catalysts*. This agreed well with results of the LC-analysis (Table II–3). The 10.9Cu *thin layer catalyst* and the 14.9Cu *thick layer catalyst* exhibited increased oxidation states of Cu metal particles compared to those Cu metal particles of catalysts obtained by the same calcination mode (compare Figure II-5).

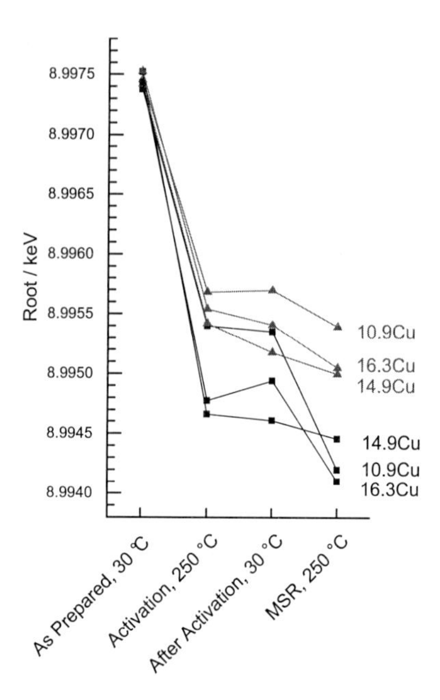

Figure II-5 Average oxidation state of Cu metal particles after certain reaction steps for *thin layer catalysts* (triangles) and *thick layer catalysts* (squares). Activation was performed as TPR by heating to 250 °C at 5 °C/min in 5 % H_2, balanced by He.

Linkage of surface atoms of Cu metal particles to oxygen atoms of the support material may cause the increased oxidation state of Cu metal particles.[127,128] Hence, the average oxidation state permitted to estimate interface between Cu metal particles and silica support. Accordingly, Cu metal particles of *thin layer catalysts* revealed distinctly larger interfaces to SBA-15 than Cu metal particles of *thick layer catalysts*. Moreover, the methanol containing feed during MSR may lead to formation of Cu surfaces which might be enriched with oxygen atoms.[129] In addition, reaction of water with strained Cu nanoparticles may form similar Cu surfaces.[128] Furthermore, adsorbed molecule, *i. e.* methoxy[130] or formate or methyl formate[131], may result in oxidized Cu atoms on Cu metal particle surfaces[132]. Partly oxidized Cu metal particles supported on SBA-15 were still present after activation and reduction continued during MSR. Furthermore, the apparent oxidation state correlated with microstrain or Cu particle sizes (see below in XAS *II.3.5* and XRD *II.3.6*).

II.3.4 Effect of Cu particle size on Cu K edge XANES profile

XANES spectra of *thin layer catalysts* and *thick layer catalysts* and reference Cu foil are compared in Figure II-6. XANES spectra of Cu/SBA-15 catalysts measured under MSR conditions differed from that of reference Cu foil.

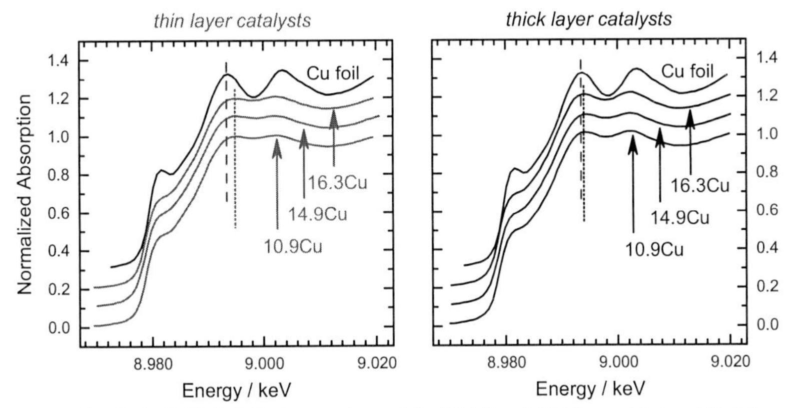

Figure II-6 Cu K edge XANES of *thin layer catalysts* and Cu foil, left, and *thick layer catalysts* and Cu foil, right. Cu/SBA-15 XANES were measured during MSR in 2 % MeOH and 2 % H_2O at 250 °C. Broken line indicates absorption feature at 8.940 keV of Cu foil. Dotted lines mark discussed maxima of XANES of Cu/SBA-15 catalysts. Spectra are shown with an offset of 0.1.

The amount of consumed H_2 during activation indicated complete reduction of CuO_x particles to Cu metal particles present in activated catalysts (Figure 0-6 in *Appendix*). In XANES spectra of Cu/SBA-15 catalysts the double peak around 9 keV was damped compared to the double peak in the XANES spectrum of a reference Cu foil. Moreover, XANES spectra of *thick layer catalysts* showed a more resolved double peak than XANES spectra of *thin layer catalysts*. The peak maxima at 8.987 keV was observed at higher energies in *thin layer catalysts*. This energy shift was hardly observed in XANES spectra of *thick layer catalysts* (see Figure II-6 dotted and broken lines). Therefore, shifts of peak maxima indicating also partial oxidation of Cu metal particles were observed although the XANES profiles corresponded to that of Cu metal foil. The amplitude of the double peak in Cu XANES strongly depends on Cu particle size.[36] Simulation of Cu K edge XANES resulted in less pronounced peaks in the XANES of Cu clusters consisting of 55 Cu atoms and finally appeared as one peak in the XANES of a cluster comprising 13 Cu atoms. Thus, Cu metal particles in *thin layer catalysts* consisted of more than 55 Cu atoms, and Cu metal particles in *thick layer catalysts* showed even further enlarged Cu metal particles.

II.3.5 EXAFS of Cu/SBA-15 catalysts at Cu K edge after TPR during MSR

Cu metal particles present in Cu/SBA-15 catalysts during methanol steam reforming were found by XANES and XRD (see below). Moreover, the pseudo radial distribution functions of Cu/SBA-15 catalysts showed characteristics similar to that of Cu metal foil. The FT($\chi(k)\cdot k^3$) of *thin layer catalysts* (green dots) and *thick layer catalysts* (black line) possessing the same Cu content are depicted in Figure II-7. All catalysts showed nearly the same peak positions but differed in amplitude. The major peak at around 2.1 Å corresponded to the first Cu neighbors. Following overlapping peaks represented scattering of more distant Cu atoms or multiple scattering. The different amplitudes indicated different particle sizes or different static disorder in Cu nanoparticles. Compared to *thick layer catalysts*, *thin layer catalysts* showed distinctly smaller amplitudes and thus smaller Cu metal particles or increased static disorder. This agreed well with results of TPR (*II.3.1*). Interestingly, the difference of amplitudes did not correlate with Cu contents. The differently calcined 14.9Cu catalysts showed the smallest deviation in amplitudes followed by 10.9Cu and 16.3Cu catalysts.

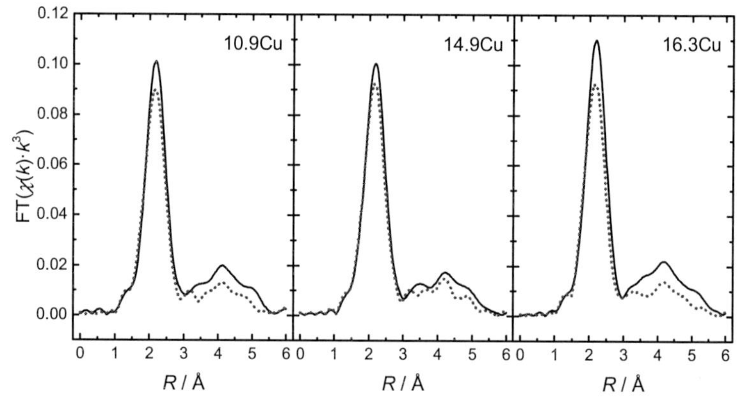

Figure II-7 Phase shifted FT($\chi(k)\cdot k^3$) of Cu/SBA-15 catalysts during MSR in 2 % MeOH and 2 % H_2O at 250 °C. Catalysts from left to right: 10.9Cu, 14.9Cu and 16.3Cu. *Thin layer catalysts* are depicted as green dots and *thick layer catalysts* are depicted as black lines.

The little shoulder in the left flank of the first peak at 1.3 Å indicated some contribution of oxygen atoms belonging possibly to the support material. The Cu metal particles were expected to be fixed to the support material. Therefore, the pseudo radial distribution function was simulated on the basis of the Cu structure including a distance from Cu to silica surface oxygen atoms. The weighting of the contribution of the Cu-O scattering path derived from the average oxidation states and resulted in clearly improved fits for *thin layer catalysts*. The model included four Cu-Cu single scattering paths up to a distance of 5.112 Å, and one additional Cu-O single scattering path at a distance of 1.85 Å for *thin layer catalysts*. Considering multiple scattering paths yields more appropriate theoretical XAS spectra.[133] Here, four triangular multiple scattering paths with a length of 4.3 to 4.8 Å and three linear multiple scattering paths with a length of 5.112 Å were considered. The FEFF7[134] software package was used to calculate the scattering paths. Independence of fit parameters was ensured performing a F-Test. Confidence intervals were calculated assuming 95 % certainty. Several constraints were necessary to refine all parameters independently. In detail, E_0 of all paths was correlated to be the same. Coordination numbers (*CN*s) were kept invariant except the first Cu-Cu scattering path, which was allowed to vary due to reduced Cu metal particle sizes. Quality of XAS spectra did not allow reliable refinement of *CN*s of other scattering paths. The reduced Cu metal particle size was thus considered in the Debye-Waller-Factors (DWFs). The distances R between atoms were refined for all single scattering paths. Rs of the multiple scattering paths were calculated from corresponding distances of single scattering paths. Debye-Waller-Factor (DWF), σ^2, of the first Cu-Cu single scattering path and the Cu-O single scattering path were fitted independently. Other DWFs of Cu-Cu single scattering paths were correlated to be equal. Additionally, DWFs of triangular multiple scattering paths were also correlated to be equal. The same constraints were applied to DWFs of the linear multiple scattering paths. This resulted in ten independent parameters. The measured FT($\chi(k)\cdot k^3$) (line) and the theoretical FT($\chi(k)\cdot k^3$) (dots) of 16.3Cu *thin* and *thick layer catalysts* at 250 °C in MSR feed are illustrated in Figure II-8. The measured and the theoretical magnitude and imaginary parts were in good agreement for both differently calcined catalysts in the fitting range of 1 Å ≤ R ≤ 5.2 Å.

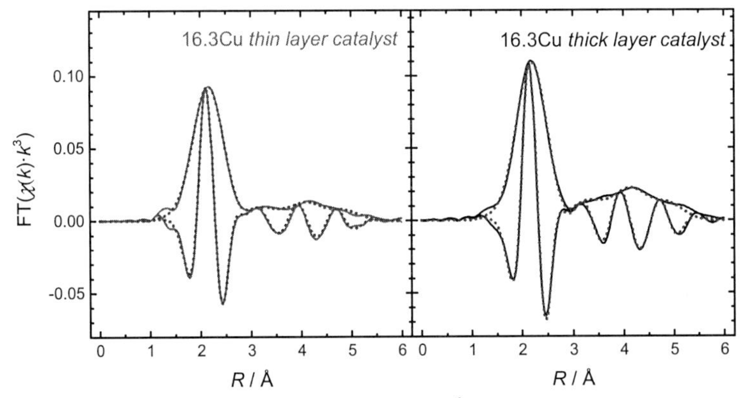

Figure II-8 Experimental (line) and refined (blue dots) FT($\chi(k)\cdot k^3$) and the corresponding imaginary parts. 16.3Cu *thin layer catalyst* and *thick layer catalyst* is plotted in the left and the right diagram, respectively.

Fit results are given in Table II–4. In comparison to Cu foil, all catalysts showed decreased *CN*s indicating small Cu metal particles present on SBA-15. Cu atoms on particle surfaces possess incomplete coordination spheres and thus exhibit distinct reduced *CN*s. Due to inharmonic vibration of surface atoms, observed *CN*s are further decreased.[135] This temperature dependent effect[119] leads to estimation of apparently larger *CN*s[136]. By refining DWFs simultaneously, the overestimation may be completely avoided for temperature lowers than 7 °C and diminished at higher temperatures.[137]

Cu metal particles were particularly small in *thin layer catalysts*. Cu metal particles of 14.9Cu *thin layer catalyst* (25 Å) were enlarged in comparison to nearly equally sized Cu metal particles in10.9Cu (18 Å) and 16.3Cu (20 Å) *thin layer catalysts*. Cu metal particles in *thick layer catalysts* were larger and particle sizes increased with increasing Cu content (Table II–4). Although the particle sizes estimated from refined *CN*s were inaccurate and showed large errors, the different sizes of Cu metal particles of *thin* and *thick layer catalysts* were in good agreement with results from XRD (*II.3.6*) and TPR (*II.3.1*).

Inharmonic vibration occurring in small Cu metal particles result in apparently reduced Cu-Cu distances in the refinement.[119] Reduced first shell distances were also observed in Cu metal particles supported on carbon.[138] Distances of about 2.49 Å in *thin layer catalysts* and 2.51 Å in *thick layer catalysts* were distinctly smaller than 2.54 Å found in Cu foil. These distances between Cu atoms observed in *thin layer catalysts* corresponded to Cu metal particles exhibiting 15 Å in diameter.[138] This agreed well with the estimated particle sizes of Cu metal particles in *thin layer catalysts* derived from *CN*s (Table II–4).

Table II–4 EXAFS fit results of catalysts in a stream of 2 % MeOH and 2 % H_2O at 250 °C. Structural parameters were refined in the range of 1 Å ≤ R ≤ 5.2 Å. Simultaneously measured Cu foil at 30 °C served as Cu reference. Uncertainty of last digit is given in round brackets.

	thin layer catalysts			*thick layer catalysts*			reference
	10.9Cu	14.9Cu	16.3Cu	10.9Cu	14.9Cu	16.3Cu	Cu
$CN_{Cu\text{-}Cu}$	8.3(6)	9.0(6)	8.4(5)	8.7(5)	9.4(5)	9.8(4)	10.5(5)
$d_{particle}$ / Å[137]	23 ± 8	28 ± 8	25 ± 5	27 ± 5	35 ± 6	38 ± 5	-
$R_{Cu\text{-}Cu}$ / Å	2.490(4)	2.490(4)	2.490(3)	2.511(2)	2.505(2)	2.511(2)	2.540(1)
$\sigma^2_{Cu\text{-}Cu}$ / Å^2	0.0160(7)	0.0165(7)	0.0157(6)	0.0153(5)	0.01613(5)	0.0155(4)	0.0077(3)
σ^2_{linear} / Å^2	0.034(2)	0.035(2)	0.032(1)	0.0265(5)	0.0273(5)	0.0260(4)	0.0146(3)

Debye-Waller-Factors (DWFs) comprise static and thermal disorders. Since all catalysts were measured at the same temperature, changes of DWFs may be interpreted as changes in the static disorder of Cu metal particles.[10] Increased DWFs indicated more disordered Cu metal particles. Longer Cu-Cu backscattering paths are more sensitive towards static disorder. Therefore, the DWFs of linear multiple scattering paths were chosen and listed in Table II–4. Once again, *thin layer catalysts* revealed less ordered Cu metal particles ($\sigma^2 \approx 0.033$), while the decreased DWFs of *thick layer catalysts* ($\sigma^2 \approx 0.027$) indicated more ordered Cu metal particles.

Just as CuO_x particles of *thin layer precursors* being small and disordered, the *thin layer catalysts* showed small Cu metal particles exhibiting pronounced structural disorder. The larger CuO_x particles of *thick layer precursors* formed larger and more ordered Cu metal particles supported on SBA-15 in *thick layer catalysts*. The disorder of Cu metal particles was apparently size related. The size and the disorder of Cu metal particles corresponded well to their preliminary CuO_x particles in oxidic precursor. During the transformation of oxidic precursors to Cu/SBA-15 catalysts structural characteristics of supported particles were preserved. Therefore, the concept of *chemical memory effect*[18,45,98] can be applied on Cu/SBA-15 catalysts.

II.3.6 XRD of Cu/SBA-15 catalysts during activation and MSR

XRD pattern were measured during activation, after activation, and during methanol steam reforming. After reaching 250 °C during activation Cu metal particles were found on SBA-15. XRD pattern showed solely peaks of Cu metal phase and no further signals of copper oxides. XRD pattern of all catalysts under working conditions, *i. e.* 2 % H_2O and 2 % MeOH at 250 °C, are depicted in Figure II-9. Measured data of *thin layer catalysts* and *thick layer* catalysts are illustrated in Figure II-9 left and right, respectively. According to the synthesis mode, XRD pattern were divided into two groups. First, *thin layer catalysts* showed Cu(111) diffraction peaks at around 43 °2θ. A shoulder, representing the Cu(200) diffraction peak, became detectable with increasing Cu content. However, both peaks overlapped and showed bell shaped profiles. Second, *thick layer precursors* showed tapered Cu(111) and Cu(200) diffraction peaks of Cu metal phase. These tapered peaks still overlapped. With increasing Cu content these peaks became better resolved. Hence, the difference in peak profiles of XRD pattern depended on both, calcination mode, and Cu content. XRD patterns of 10.9Cu *thin* and *thick layer catalysts* were nearly congruent, while XRD patterns of 27.1Cu *thin layer catalyst* and 27.1Cu *thick layer catalyst* were clearly different. The various peak profiles of differently calcined catalysts are discernable in Figure II-9. Moreover, broadening and flattening of diffraction peaks were caused by diminished crystallite size.[24] The bell shaped Cu diffraction peaks of *thin layer catalysts* similar to a Gaussian function profile indicated microstrain in Cu metal particles. Strained Cu metal particles were previously observed in activated Cu/ZnO[12,19] and Cu/ZnO/Al_2O_3 catalysts[10]. More sharp diffraction peaks of *thick layer precursors*, which were fitted sufficiently without any contribution of Gaussian functions, indicated regularly structured Cu metal particles. In order to disentangle crystallite size and microstrain of Cu metal particles, pseudo Voigt functions (pVfs) were used to refine the measured pattern. According to de Keijser *et al.*[117] Gaussian contributions of pVfs and thus microstrain in Cu metal particles were calculated. This analysis was performed during activation at 250 °C, and after activation in H_2 atmosphere as well as during MSR at 250 °C. Results are summarized in Figure II-10 together with corresponding crystallite.[116]

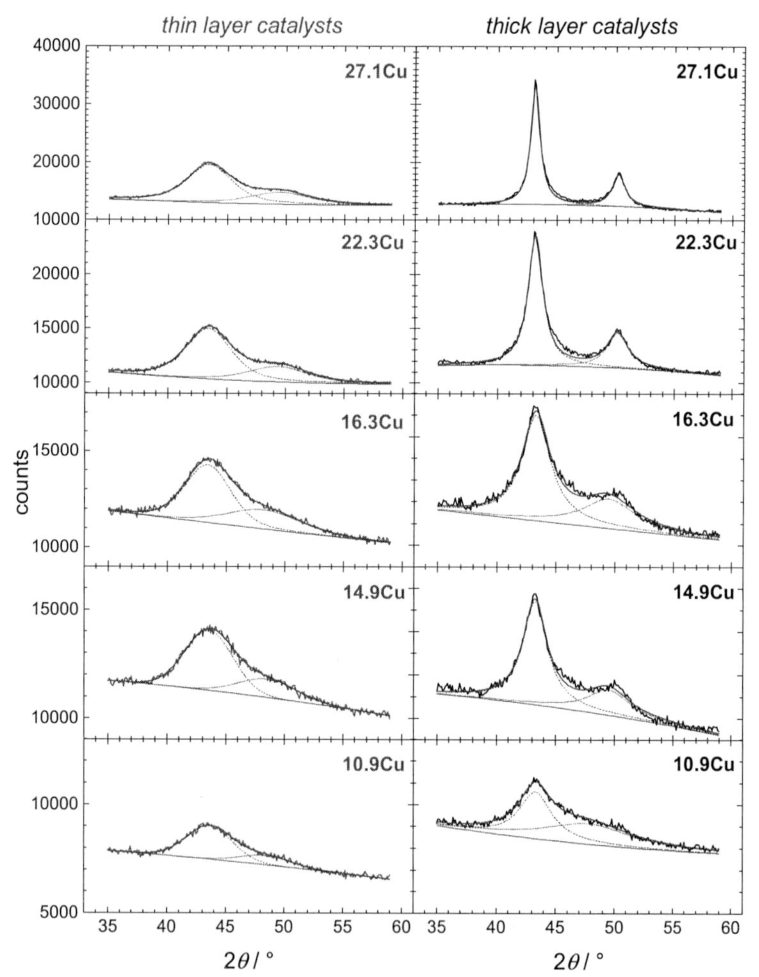

Figure II-9 XRD pattern measured at 250 °C during methanol steam reforming in 2 % MeOH and 2 % H_2O balanced by He of *thin layer catalysts* (left, green) and *thick layer catalysts* (right, black). Cu loading increases from bottom to top. Blue lines represent the entire fit and the estimated background. Blue dashed lines represent Cu(111) diffraction peak at 43 °2θ and blue dots represent Cu(200) diffraction peak at 50 °2θ.

In Figure II-10 left and right results are shown for *thin layer catalysts* and for *thick layer catalysts*, respectively. Filled symbols represent microstrain and empty symbols represent crystallite sizes. *Thin layer catalysts* consisted of Cu metal particles of approximately 2 nm independent of Cu loading. These Cu metal particles showed a highly disordered structure indicated by microstrain in the range from 1.7 to 3.4 %. Errors were estimated to be in the range of 0.5 % to 1 % of microstrain. These great uncertainties were mainly caused by inadequate data quality and can be seen at the varying microstrain of Cu metal particles in 14.9Cu and 16.3Cu *thin layer catalysts*. Hence, similar microstrain was assumed for Cu metal particles of *thin layer catalysts*.

Microstrain and size of Cu metal particles of *thick layer catalysts* were different compared to those of *thin layer catalysts* with the same Cu content. Deviations in microstrain of *thick layer catalysts* were so small that data points are overlapping (Figure II-10, fulfilled squares, right). Hence, the structure of Cu metal particles of *thick layer catalysts* was more ordered, and contained fewer defects. Mean particle sizes were estimated to be about 2-3 nm for 10.9Cu, 14.9Cu, and 16.3Cu ranging up to 5 nm in 22.3Cu and 8 nm in 27.1 nm *thick layer catalysts*. A correlation between Cu content and particle size was found for higher Cu loadings. Cu metal particle sizes of here reported Cu/SBA-15 catalysts corresponded well to Cu metal particles sizes of various Cu catalysts reported in literature. Cu metal particle sizes were found to be between 8-20 nm in Cu/ZnO[19], and 5.1-7.7 nm in Cu/ZnO/Al_2O_3[10] catalysts. Cu metal particles supported on silica exhibited particles size from 0.7 nm for 5.0 wt.% Cu/SBA-15 catalysts derived by precursor decomposition in He atmosphere[54], 5 nm for 10 wt.% Cu/SiO_2 catalysts, and 34 nm for 30 wt.% Cu/SiO_2 catalysts[16].

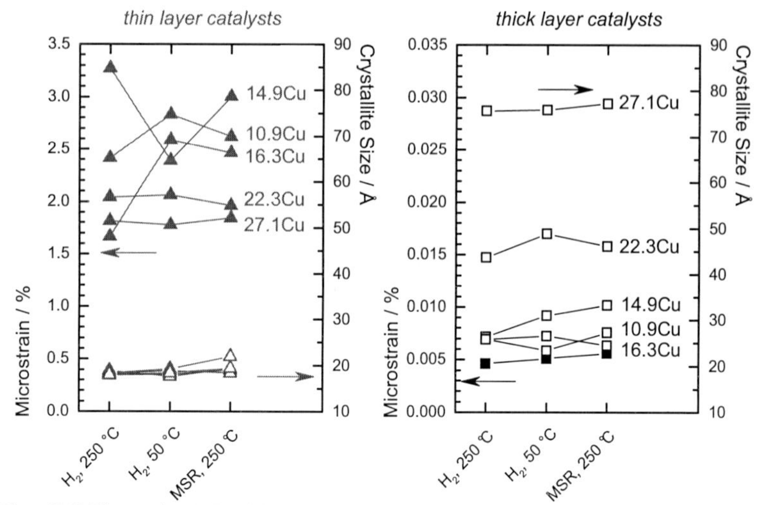

Figure II-10 Microstrain and size of Cu metal particles, derived from X-ray diffraction peak profile analysis of *thin layer catalysts* (left, triangles) and *thick layer catalysts* (right, squares). Fulfilled symbols represent microstrain and empty symbols represent crystallite sizes.

The Cu metal particles supported on SBA-15 were smaller than the diameter of the mesopores. Hence, it is assumed that Cu metal particles were mainly present in mesopores. Especially, Cu metal particles of *thin layer catalysts* might even fill up micropores. The size of Cu metal particles of *thin layer catalysts* corresponded to the defined maximum of micropore diameter of 2 nm. The Cu metal particles of Cu/SBA-15 catalysts were mainly invariant during activation and methanol steam reforming with regard to their microstrain and their size. Slight sintering of Cu metal particles was indicated in 27.1Cu *thin layer catalyst* as well as 14.9Cu and 27.1Cu *thick layer catalysts* (Figure II-10). Hence, the high surface area of SBA-15 stabilized the Cu metal particles formed during activation.

II.3.7 Cu surface areas of Cu/SBA-15 catalysts

Surface areas of Cu metal particles were estimated from quantities of chemisorbed oxygen. The intrinsic surface areas of Cu nanoparticles, derived from freshly reduced oxidic precursors and the corresponding dispersion of Cu nanoparticles are depicted in Figure II-11 on the left. In the following it is assumed that these Cu surface areas are equal to the amount of active sites in corresponding Cu/SBA-15 catalysts to calculate turn over frequencies (TOFs) (see *II.3.8*).

Two trends in the evolution of Cu surface areas were observed for freshly reduced Cu/SBA-15 catalysts: First, *thin layer catalysts* formed more dispersed Cu metal particles than the corresponding *thick layer catalysts*. Second, higher Cu loadings in the oxidic precursors led to less dispersed Cu nanoparticles supported on SBA-15. A nearly linear decrease in dispersion with increasing Cu content was found for Cu metal particles obtained from *thick layer precursors*. The dispersion of Cu metal particles obtained from *thin layer precursors* gave a more complex picture.

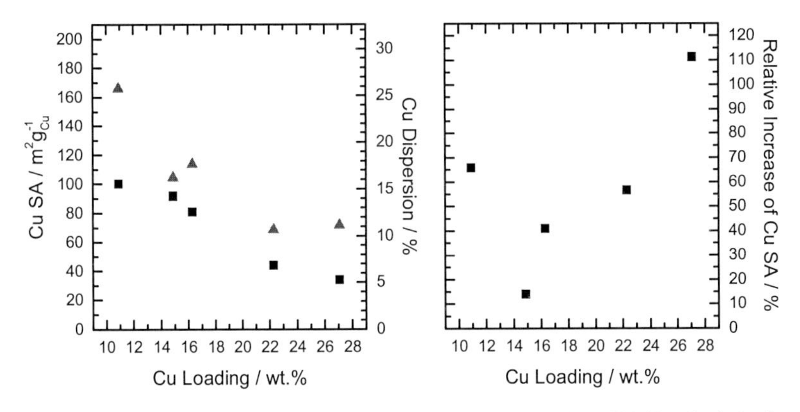

Figure II-11 Intrinsic Cu surface area (SA) of Cu metal particles supported on SBA-15 and calculated dispersion of Cu metal particles supported on SBA-15 on the left, after activation of oxidic precursors in 5 % H_2 for 2 h at 250 °C. Relative increase in Cu surface areas referred to *thick layer catalysts* on the right.

The increase of Cu surface area as a consequence of calcination mode showed no systematical trend. For 10.9Cu, 16.3Cu, and 22.3Cu an increase between 40 to 70 % was measured (Figure II-11, right). However, 14.9Cu catalysts showed only 10 % increase and 27.1Cu catalysts showed 110 % increase in Cu surface area. As indicated by XRD, 27.1Cu *thin layer precursors* showed no long range order and therefore well dispersed CuO_x particles while *thick layer precursor* exhibited small but detectable CuO_x particles supported on SBA-15. It was in good agreement that Cu metal particles obtained from well dispersed CuO_x particles exhibited larger intrinsic surface areas than Cu metal particles obtained from larger CuO_x particles.

Another explanation for differences in Cu surface areas bases on a report of van der Grift *et al.*.[120] They concluded, that the amount of flat coherent Cu particle faces with less steps, less holes, and less kinks increased the measured Cu surface areas. Oxygen atoms preferably chemisorb on flat faces. Accordingly, Cu metal particles obtained from *thick layer precursors* might exhibit small faces, which were interrupted by steps and kinks forming a spherical particle. Cu metal particles obtained from *thin layer precursors* might exhibit larger faces and fewer steps indicating flat Cu metal particles with enriched contribution of continuous faces.

Considering the structure of CuO_x particles in the oxidic precursors, both, dispersion of Cu metal particles as well as different shapes of Cu metal particles after reduction of oxidic precursors may affect the measured differences of Cu surface areas.[123] It is possible, that well dispersed CuO_x particles of *thin layer precursors* transformed during reduction to flat Cu metal particles anchored on the SBA-15 mesopore walls. But it is also possible, that many small Cu metal particles were formed. The larger CuO_x particles of *thick layer precursors* yielded more spherical Cu metal particles, which were consequently not as fine distributed on the support as those of *thin layer catalysts*.

II.3.8 Methanol steam reforming over Cu/SBA-15 catalysts

Cu nanoparticles of Cu/SBA-15 catalysts converted methanol and water with high selectivity to H_2 and CO_2 under the employed conditions. Traces of side products formaldehyde and methyl formate were identified (Figure 0-7) in gas chromatograms as reported earlier[13] for Cu/SiO_2 catalysts. CO was not observed. Short contact times[139] and conversion below 20 % suppressed CO formation derived from the reversed water gas shift reaction[140,141]. Figure II-12 illustrates the development of the H_2 formation rates for *thin layer catalysts* on the left and *thick layer catalysts* on the right. The performances of catalysts having identical composition were different with respect to calcination mode. Furthermore, the performance of catalysts varied with varying Cu content while same calcination mode was employed.

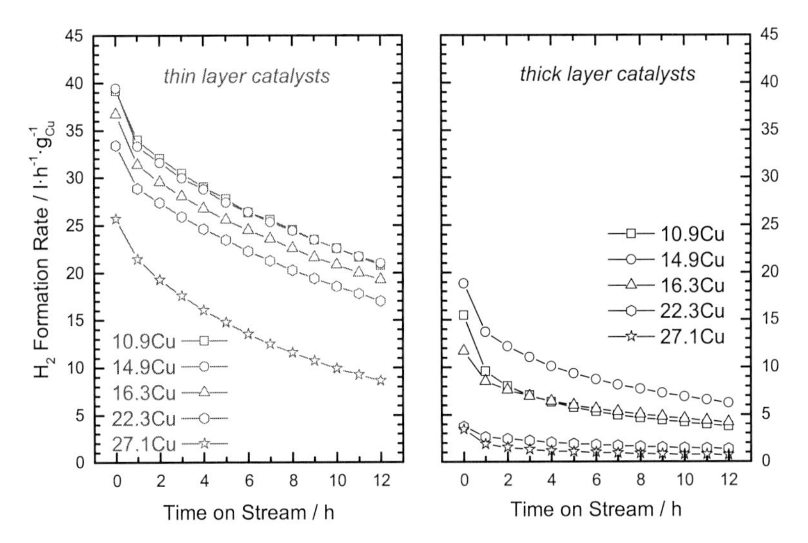

Figure II-12 H_2 formation rates during methanol steam reforming in 2 % MeOH and 2 % H_2O at 250 °C over Cu/SBA-15 catalysts for 12 h time on stream. Activities of *thin layer catalysts* are presented in the left and those of *thick layer catalysts* are presented in the right diagram.

Thin layer catalysts showed higher H_2 formation rates than *thick layer catalysts* exhibiting the same Cu content. In addition, t*hin layer catalysts* exhibited reduced H_2 formation rates with increasing Cu content. The activity of 10.9Cu *thin layer catalyst* was almost the same as the activity of 14.9Cu followed by 16.3Cu and then by 22.3Cu *thin layer catalysts*. 27.1Cu *thin layer catalyst* revealed distinctly reduced catalytic activity. A more complicated order of activity was found for *thick layer catalysts*. 14.9 Cu *thick layer catalysts* showed the highest H_2 formation rate followed by 16.3Cu and 10.9Cu *thick layer catalysts*, which produced nearly equal amounts of H_2. Then, lower H_2 formation rates of 22.3Cu and further reduced H_2 formation rates of 27.1Cu *thick layer catalyst* were observed.

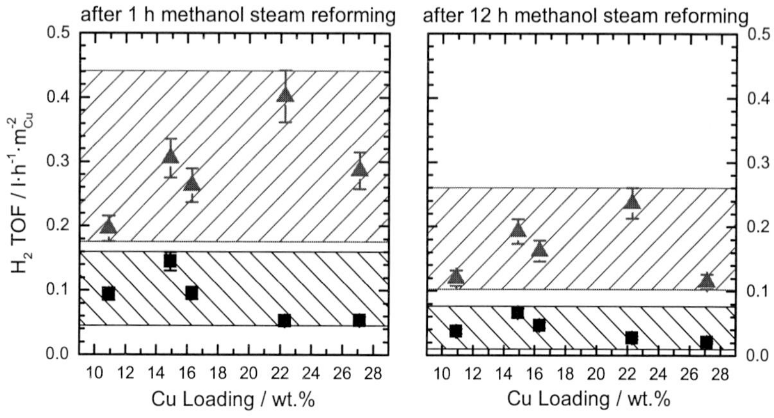

Figure II-13 H_2 TOFs of *thin layer catalysts* (triangles) and *thick layer catalysts* (squares). Methanol steam reforming proceeded for 1 h (left) and for 12 h (right) time on stream in 2 % MeOH and 2 % H_2O at 250 °C.

Catalytic activity could depend on available Cu surface area as it is commonly discussed.[5] Besides Cu surface areas, particular microstructural characteristics[12,46] or Cu particle shape[120] may affect the H_2 production by MSR. Therefore, the intrinsic catalytic activity of Cu surface atoms was calculated as turn over frequency (TOF). The H_2 formation rate was related to the estimated Cu surface areas. Although Cu surface measurements are prone to errors, the measured Cu surface areas were assumed as sufficiently accurate. Applying the same setup and adjusting the same conditions reduced errors and enabled comparison of Cu surface areas of different Cu/SBA-15 catalysts. TOFs of *thin layer catalysts* and *thick layer catalysts* are depicted in Figure II-13 as triangles and squares, respectively. On the left, the H_2 TOFs are shown after 1 h time on stream, and on the right, H_2 TOFs are shown after 12 h time on stream. The given error bars were estimated at 10 % due to inaccuracy of scales, of chromatograms, of experimental setup, and deviations of Cu surface areas. Differently shaded areas distinguish the range of H_2 TOFs of *thin layer catalysts* on the top and *thick layer catalysts* on the bottom. The intrinsic activity was generally higher than that of *thick layer catalysts* independent of Cu loading.

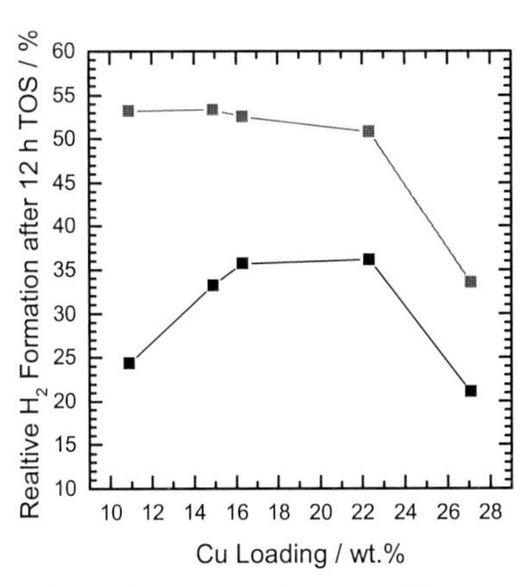

Figure II-14 Remained H_2 TOF of *thin layer catalysts* (triangle) and *thick layer catalysts* after 12 h time on stream (TOS) in 2 % MeOH and 2 % H_2O at 250 °C.

During 12 h time on stream the catalysts deactivated. Nearly the same trend of H_2 TOFs of Cu/SBA-15 catalyst was found after 1 h and after 12 h time on stream (Figure II-14). Only the spreading of absolute TOFs was reduced. This indicated similar deactivation rates of Cu/SBA-15 catalysts of each sample set according to calcination mode. *Thin layer catalysts* showed not only higher TOFs compared to *thick layer catalysts*, but also showed less reduced TOFs after 12 h methanol steam reforming (Figure II-14). T*hin layer catalyst* exhibited after 12 h time on stream H_2 TOF in methanol steam reforming between 52 % and 54 % compared to H_2 TOFs after 1 h time on steam. Only the 27.1Cu *thin layer catalyst* showed further reduced H_2 TOF at 35 % after 12 h time on stream.

Thick layer catalysts 14.9Cu, 16.3Cu, and 22.3Cu showed reduced H_2 TOFs between 33 % and 35 % after 12 h time on stream. Further deactivation was found for 10.9Cu *thick layer catalyst* (24 % remained activity) and 27.1Cu *thick layer catalyst* (22 % remained activity).

II.3.9 Size of Cu metal particles supported on SBA-15

Activation of oxidic precursors was achieved by reduction of CuO_x particles to Cu metal particles in 5 % H_2 atmosphere at 250 °C. This reduction proceeded via a Cu_2O like intermediate similar as reported for CuO/ZnO systems.[11,97] The Cu particle sizes of Cu/SBA-15 catalysts obtained from different analytical technics are compared in Table II–5. The Cu metal particles (XAS) and the Cu crystallite sizes (XRD) were smaller than 8 nm. Crystallite sizes and particle sizes can be presumed to be equal for nanoparticles smaller than 20 nm in diameter.[142] XRD analysis may yield an apparently increased size for nanoparticles, because only particles exceeding sufficient long range order are detectable. XAS detects all available particles independent of their size. However, the refined *CN*s used for calculation of XAS based particle sizes tend to yield apparently smaller particles due to inadequate data quality. The refinement of *CN* is prone to errors due to thermal effects, defects in structure, and wide particle size distributions.[119] Therefore, the XAS based particle sizes possess large uncertainties due to error propagation. Interestingly, the XRD based particle sizes agreed well with those obtained from XAS analysis within the errors. This suggested a homogenous particle size distribution of Cu metal particles in each Cu/SBA-15 catalysts. However, the Cu particle sizes, calculated from determined Cu surface areas, were enlarged compared to Cu particle sizes derived from XRD or XAS. This may be caused by large interfaces between Cu metal particles and the support. These interfaces were not accessible for N_2O molecules, used for Cu surface determination. Hence, the smaller Cu surface pretended larger Cu metal particles. Accordingly, higher deviation between N_2O based particle sizes and XRD or XAS based particle sizes may indicate larger interfaces between Cu metal particles and SBA-15.

Table II–5 Calculated Cu particle sizes. (a) results from Scherrer equation (b) results from *CN* of EXAFS fit[137] (c) assuming spherical Cu metal particles (d) ratio between D_{N2O} and $D_{XRD/XAS}$ = (D_{XRD} + D_{XAS})/2 if possible or $D_{XRD/XAS}$ = D_{XRD}.

	D_{XRD} / nm (a)		D_{XAS} / nm (b)		D_{N2O} / nm (c)		$D_{N2O}/D_{XRD/XAS}$ (d)	
	thin layer	*thick layer*	*thin layer*	*thick layer*	*thin layer*	*thick layer*	*thin layer*	*thick layer*
10.9Cu	1.9	2.6	2.3 ± 0.8	2.7 ± 0.5	4.1	6.7	2.0	2.5
14.9Cu	1.9	3.4	2.8 ± 0.8	3.5 ± 0.6	6.4	7.3	2.7	2.1
16.3Cu	1.9	2.4	2.5 ± 0.5	3.8 ± 0.5	5.9	8.3	2.7	2.7
22.3Cu	1.9	4.9	-	-	9.8	15.3	5.2	3.1
27.1Cu	2.1	7.5	-	-	9.4	19.9	4.5	2.7

The sizes of CuO_x particles in the oxidic precursors were preserved in Cu metal particles on SBA-15 formed during activation and MSR. Well dispersed and amorphous CuO_x particles of *thin layer precursors* were transformed into well dispersed Cu metal particles in *thin layer catalysts*. The particles were about 2 nm in all *thin layer catalysts* (XAS, XRD). Reduction of less dispersed CuO_x particles of *thick layer precursors* resulted in larger Cu metal particles in *thick layer catalysts*. The diameters of these Cu metal particles were located in the range of 2.4 to 7.5 nm (XAS, XRD). Activation of both *thin layer* and *thick layer precursors* led to Cu metal particles exhibiting smaller diameters than the mesopore diameter of SBA-15 (9 nm[58]). Hence it was likely that Cu metal particles were in the mesopores as previously reported.[50]

The Cu metal particles of *thin layer catalysts* would fit the diameter of choked micropores (both around 2 nm). Therefore, as proposed in *Chapter I* micropores may be filled by CuO_x particles. Formation of Cu metal particles may proceed without distinct migration of Cu atoms. Hence, the freshly formed Cu metal particles of *thin layer catalysts* may be located in intrawall micropores. Thus, the Cu metal particles showed a large interface between Cu metal particles and SBA-15 pore walls. The ratio of estimated particle size derived from Cu surface measurements and Cu particle size derived from XRD and XAS served as an expression for the Cu/SiO_2 interface (see Table II–5). The *thin layer catalysts* showed an increasing ratio $D_{N2O}/D_{XRD/XAS}$ with increasing Cu content. Therefore, the interface between Cu metal particles and SBA-15 increased with increasing Cu loading. That was likely the reason for the large Cu surface areas in *thin layer catalysts*. The well dispersed Cu metal particles were stabilized by large interfaces between Cu metal particles and SBA-15 preventing them from sintering.

Conversely, *thick layer catalysts* revealed similar interface areas between Cu metal particles and SBA-15 independent of Cu content. This indicated that *thick layer catalysts* contained more spherical shaped Cu metal particles. With increasing Cu loading larger Cu metal particles were observed exhibiting similar interfaces to SBA-15. According to a *chemical memory effect*, the spherically shaped CuO_x particles of *thick layer precursor* were transformed into spherical Cu metal particles supported on SBA-15 (compare *I.3.15*).

II.3.10 Microstrain of Cu metal particles supported on SBA-15

A further characteristic difference between *thin layer* and *thick layer catalysts* was the development of microstrain observed in the supported Cu metal particles. While *thick layer catalysts* consisted of well-ordered Cu nanoparticles possessing sizes of 2.5 to 8 nm, *thin layer catalysts* revealed distinct disorder in association with Cu metal particle sizes of 2 nm. Thus, the microstrain in these Cu metal particles of *thin layer catalysts* was apparently size related. Microstrain in Cu metal particles was discussed as consequence of epitaxial particle growth on ZnO[19] or dissolution of ZnO in Cu[19] or oxygen content in Cu[143] as well as defect rich metastable Cu phase stabilized by the Cu/ZnO interface[11,19] and not completely reduced Cu[19]. These examples indicated a chemical interplay of Cu metal particles with their support or promoter as source of microstrain. For Cu/SBA-15 catalysts it can be assumed that both, the effects of dissolution of oxygen atoms and the Cu/SiO_2 interface influenced the disorder in Cu metal particles. First, SBA-15 stabilized Cu metal particles, especially by hosting Cu metal particles in micropores in *thin layer catalysts* (compare Figure II-15 left). Second, oxygen atoms or hydroxyl groups at the Cu/SiO_2 interface may be enclosed in Cu metal particles and thus evoke microstrain. Third, microstrain might be achieved by large interfaces between Cu metal particles and SBA-15. The irregular surfaces of amorphous pore walls of SBA-15 might lead to irregularly structured Cu metal particles due to distinct interaction. It has been shown that various Cu metal particle surfaces may be accessible if Cu is deposited on surfaces of amorphous SiO_2 supports.[144] Therefore, the formation of different types of Cu metal particles on the inner surfaces of SBA-15 catalysts was likely depending on the interface (Figure II-15).

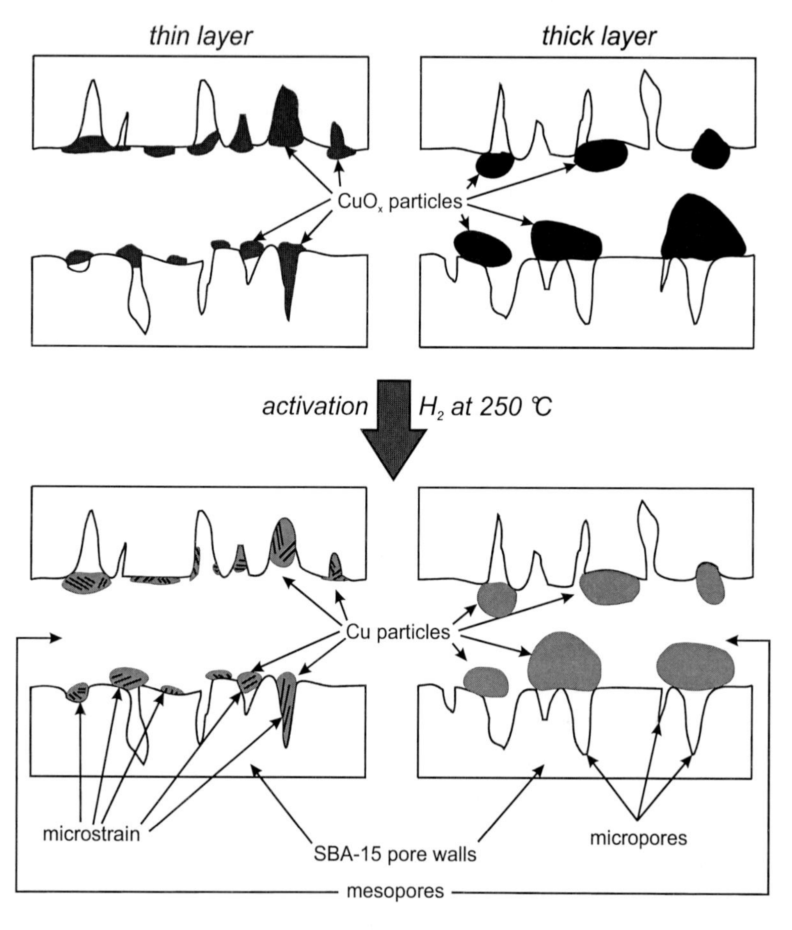

Figure II-15 Formation of different Cu/SBA-15 catalysts obtained from *thin layer precursors* and *thick layer precursors*, left and right, respectively. Smaller Cu metal particles in *thin layer catalysts* exhibit larger interface to silica pore walls and distinctive microstrain or disorder (hatching) compared to *thick layer catalysts* (no hatching).

In *thin layer catalyst* highly disordered Cu metal particles were observed. Interestingly, distinctive microstrain was found, although catalysts were completely free of ZnO. These Cu metal particles showed a large interface to SBA-15. In contrast, *thick layer catalysts* with the same chemical composition and the same support material showed less pronounced Cu/SBA-15 interfaces. In the Cu metal particles of *thick layer catalysts* little microstrain was found. These Cu metal particles exhibited an ordered Cu structure. Therefore, the Cu/SBA-15 catalysts were divided into two groups: *Thin layer catalyst* and *thick layer catalysts*, consisting of small strained Cu metal particles and larger ordered Cu metal particles, respectively. The structural characteristics of Cu metal particles in Cu/SBA-15 catalysts resembled the structural characteristics of CuO_x particles in corresponding oxidic precursors.

II.3.11 Evaluation of microstrain obtained from XRD and static disorder obtained from XAS

The detection of microstructural differences in Cu metal particles of Cu/SBA-15 catalysts during MSR at 250 °C succeeded independently using both XRD and XAS. Structural characteristics were obtained as microstrain and static disorder. The XRD based microstrain and the selected DWFs as disorder parameter derived from EXAFS analysis are depicted in Figure II-16, (f), (g), (i) for *thin layer catalysts*. These parameters showed the same evolution of microstrain and static disorder for 10.9Cu, 14.9Cu and 16.3Cu *thin layer catalysts*.

Corresponding results of *thick layer catalysts* are given in Figure II-18 (f), (g), (i). XRD profile analysis indicated little microstrain in Cu metal particles of *thick layer catalysts*. Similarly, the EXAFS refinement revealed distinctly reduced DWFs compared to *thin layer catalysts* indicating more ordered Cu metal particles. However, slight differences in DWFs among 10.9Cu, 14.9Cu, and 16.3Cu *thick layer catalysts* were found. Hence, EXAFS became more sensitive towards differences in structural disorder of sufficiently ordered Cu metal particles.

II.3.12 Structure activity correlations of thin layer catalysts

The results of structural characterization, the H_2 formation rate (a) and the TOFs (b) of *thin layer catalysts* are summarized in Figure II-16. *Thin layer catalysts* showed good performance in methanol steam reforming (MSR). With increasing loading the H_2 formation rate decreased slightly. The highest loaded 27.1Cu *thin layer catalyst* exhibited further decreased H_2 formation rate accompanied by little increase of Cu metal particle size (see Figure II-16 (h)). XRD peak profile analysis revealed that *thin layer catalysts* consisted of well dispersed Cu metal particles of similar size. Consequently, evolution of H_2 formation rates cannot be attributed to different Cu metal particle sizes. The Cu metal particles of *thin layer catalysts* exhibited also microstrain, which may influence the electronic structure of the metal particles surface and thus the catalytic activity.[99] In *thin layer catalysts* the microstrain in Cu metal particles was accompanied by increased H_2 formation rates in MSR (see (a), (f), and (g) and Figure II-17) as it was reported for Cu/ZnO[19] and Cu/ZnO/Al_2O_3[10] catalysts. The evolution of the H_2 formation rate and the evolution of the microstrain in the similarly sized Cu metal particles showed similar trends ((a) and (i)). However, the H_2 formation rates of *thin layer catalysts* may be influenced by Cu surface areas.[48] The Cu surface areas are considered in H_2 TOFs of *thin layer catalysts* (b). The trace of H_2 TOFs deviated from the trace of H_2 formation rate indicating different surfaces present on Cu metal particles in *thin layer catalysts*. The trace of H_2 TOF (b) was similar to the trace of the temperature of the maximum in H_2 consumption during activation (c) depending on Cu loading. Correlation between increased MSR selectivity towards CO_2 and thus increased H_2 formation[140] and increased reduction temperatures has been previously reported for Cu/ZnO catalysts.[11] Typically, decreased reducibility indicates larger particles. Here, this conflicted with analytical data from XAS and XRD measurements. Apparently, not only particle size influenced the reducibility but also the surface of Cu metal particles affected the reducibility and thus the catalytic activity.

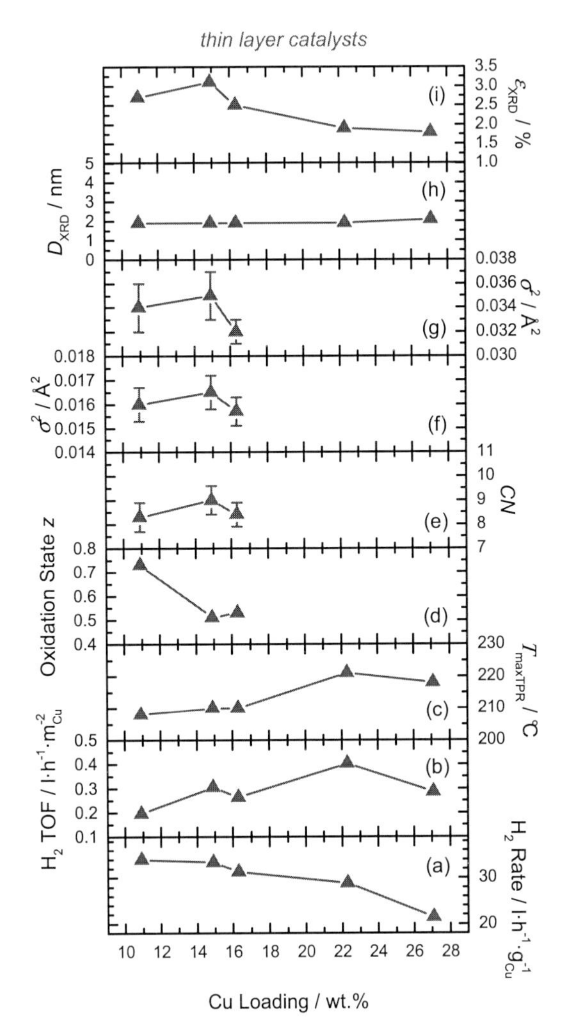

Figure II-16 Comparison of (a) H_2 formation rate (b) intrinsic activity of Cu metal particles (TOF) of differently loaded *thin layer catalysts* after 1 h time on stream with (c) peak maximum of H_2 consumption during activation, (d) average oxidation state of Cu metal particles during MSR, (e) coordination number of first Cu-Cu scattering path (f) Debye-Waller-Factor, σ^2, of first Cu-Cu scattering path, (g) Debye-Waller-Factor of linear multiple scattering paths, (h) XRD based Cu particle size, and (i) XRD based microstrain in Cu metal particles.

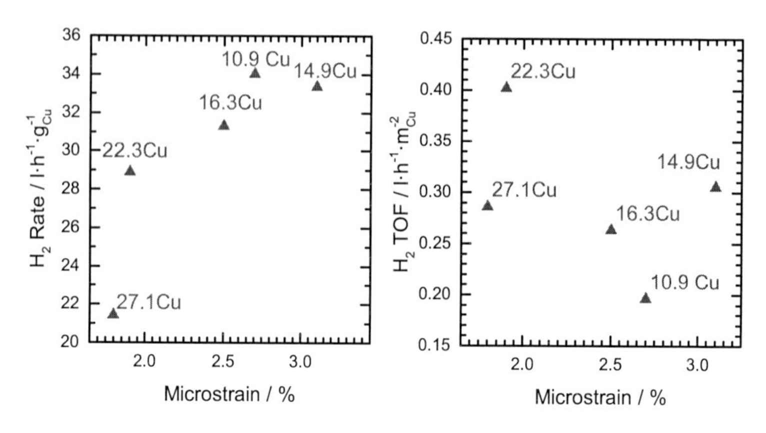

Figure II-17 Evolution of H_2 formation rates and H_2 TOFs over thin layer catalysts in methanol steam reforming depending on microstrain, left and right, respectively.

The large interface between Cu metal particles and support might evoke microstrain in Cu metal particles of *thin layer catalysts* yielding increased H_2 TOF (compare Table II–5). Probably, CuO_x particles of oxidic precursors interacted differently with the SBA-15 support. As result, different behaviors during activation were observed resulting in different properties of Cu metal particles. For instance, Cu metal particles of various shapes might be formed yielding in various Cu metal particle surfaces. Different Cu metal particle surfaces show different catalytic activities.[120] The interplay of Cu metal particle shape, Cu metal particle size, microstrain in Cu metal particles and the interface between Cu metal particles and support (Table II–5, last column) mostly affected the reducibility of the *thin layer catalysts* which correlated with the intrinsic activity.

II.3.13 Structure activity correlations of thick layer catalysts

Thick layer catalysts possessed larger Cu metal particles. The *thick layer catalysts* showed less activity in methanol steam reforming compared to *thin layer catalysts*. A compilation of MSR activity and Cu metal particle properties is given in Figure II-18 (a)-(i).

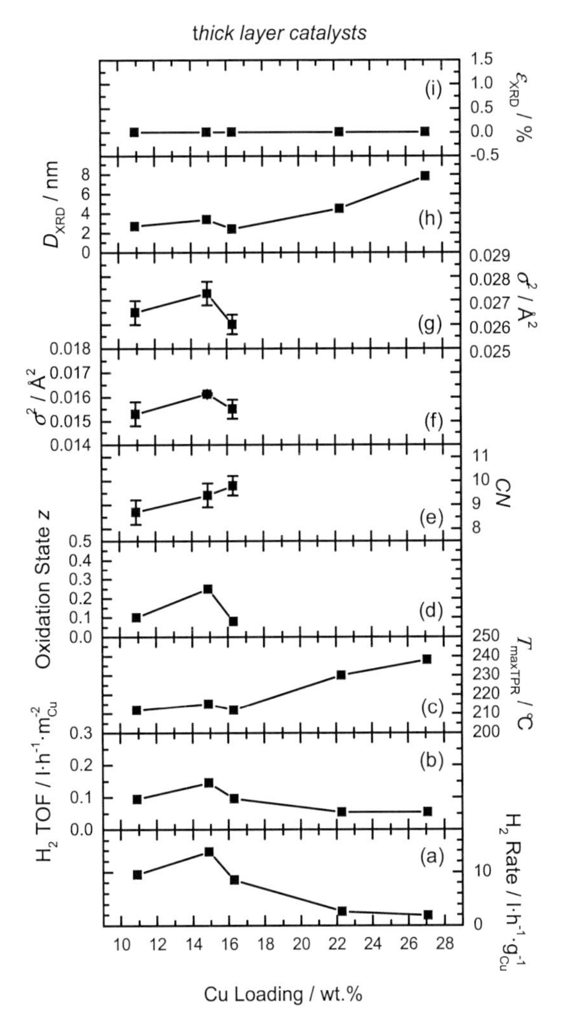

Figure II-18 Comparison of (a) H_2 formation rate (b) intrinsic activity of Cu metal particles (TOF) of differently loaded *thick layer catalysts* after 1 h time on stream with (c) peak maximum of H_2 consumption during activation, (d) average oxidation state of Cu metal particles during MSR, (e) coordination number of first Cu-Cu scattering path (f) Debye-Waller-Factor, σ^2, of first Cu-Cu scattering path, (g) Debye-Waller-Factor of linear multiple scattering paths, (h) XRD based Cu particle size, and (i) XRD based microstrain in Cu metal particles.

Cu metal particles in *thick layer catalysts* exhibited very little microstrain (i) but showed various particle sizes (h). Additionally, variation of the local static disorder was found for 10.9Cu, 14.9Cu and 16.3Cu *thick layer catalysts* ((f) and (g)). These various Cu metal particles supported on SBA-15 showed similar evolutions of the H_2 formation rate and the H_2 TOF (see (a) and (b)). The 14.9Cu t*hick layer catalyst* showed the highest H_2 formation rate and the highest H_2 TOF. 10.9Cu and 16.3Cu showed reduced H_2 formation rates and reduced H_2 TOFs. 22.3Cu and 27.1Cu showed the lowest H_2 formation rates and the lowest H_2 TOFs. The trace of temperature of the highest H_2 consumption during activation (c) correlated with the TOFs of 10.9Cu, 14.9Cu and 16.3Cu *thick layer catalysts* in a similar way as in *thin layer catalysts*. Higher loaded 22.3Cu and 27.1Cu catalysts showed a decreased reducibility with decreased H_2 TOFs. For *thick layer catalysts*, the temperature of the highest H_2 consumption correlated with the Cu metal particles sizes obtained from XRD data (g) accompanied by less activity in MSR for larger Cu metal particles.

The 14.9Cu *thick layer catalyst* exhibited a particular MSR activity. This catalyst showed the highest H_2 formation rate and the highest H_2 TOF. Furthermore, Cu metal particles of 14.9Cu *thick layer catalyst* were apparently larger than those in less active 10.9Cu and 16.3Cu *thick layer catalysts* (see (c) and (h)). Simultaneously, Cu metal particles of 14.9Cu *thick layer catalyst* revealed an increased static disorder in Cu metal particles ((f) and (g)), and indicated an increased oxidation state of Cu metal particles (e). It was reported that both affect the electronic structure of Cu metal particles[99,129,143], resulting in enhanced catalytic activity in MSR[10]. Similar to here investigated 14.9Cu *thick layer catalyst*, Günter *et al.*[11] reported of increased Cu metal particles of Cu/ZnO catalysts exhibiting increased microstrain. Additionally, increased microstrain correlated with increased H_2 TOFs in MSR.[11] The coincidentally occurring increased oxidation state and increased microstrain in Cu metal particles of 14.9Cu *thick layer catalyst* was described similarly for Cu. Böttger *et al.*[129] concluded that the oxygen atoms were enclosed in the Cu structure, evoking microstrain. Further investigations on Cu/SiO_2[16] catalysts showed that the increased average oxidation state of Cu metal particles led to suppression of undesired reversed water gas shift reaction and thus to apparently higher H_2 TOFs. The here found increased activity of 14.9Cu *thick layer catalyst* can therefore be explained by the interplay of the high static disorder in Cu metal particles (Figure II-19) and the apparent increased oxidation state (Figure II-18 (a) (b) (e)). Moreover, the H_2 formation rate correlated with the static disorder in Cu metal particles of 10.9Cu, 14.9Cu and 16.3Cu *thick layer catalysts* (compare Figure II-19). With increasing disorder parameter σ^2 of linear Cu-Cu multiple scattering paths in Cu metal particles, the H_2 formation rate increased supporting the previously found structure activity correlations of Cu/ZnO and $Cu/ZnO/Al_2O_3$ catalysts.[10]

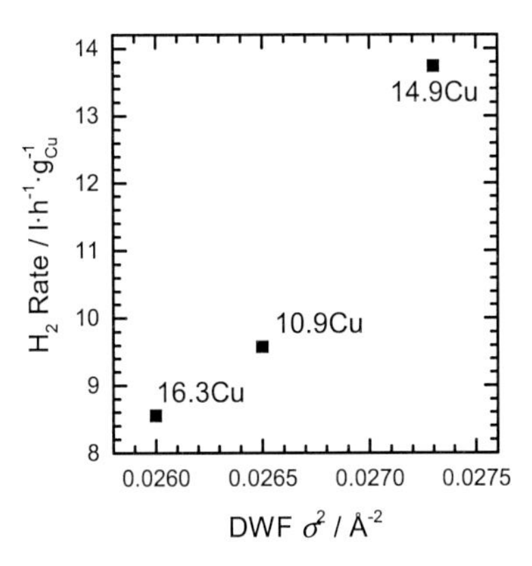

Figure II-19 H_2 formation rate over 10.9Cu, 14.9Cu, and 16.3Cu *thick layer catalysts* in dependence on static disorder parameter Debye-Waller-Factor (DWF).

Interestingly, the evolution of H_2 TOF rate correlated with the H_2 formation rate (Figure II-22). Hence, the increasing H_2 formation rate could be attributed to the various Cu metal particle sizes yielding various Cu surface areas. Additionally, the Cu surface atoms became more active, which correlated with observed disorder in Cu metal particles.[99]

II.3.14 Relationships between Cu particle size and disorder in Cu particles and catalytic activity in methanol steam reforming

Independent of calcination mode of Cu/SBA-15 catalysts the following relationships between the size of Cu metal particles and H_2 formation rate and H_2 TOF in MSR were found (Figure II-20). Larger Cu metal particles showed less H_2 formation rates and less intrinsic activity, while smaller Cu metal particles showed increased H_2 formation rates and increased intrinsic activity in MSR. Catalytic properties of solid materials depend on their particle size.[102] Cu metal particles being smaller than 3 nm showed an increased MSR activity indicating changes in electronic structure due to quantum confinement effects.[145]

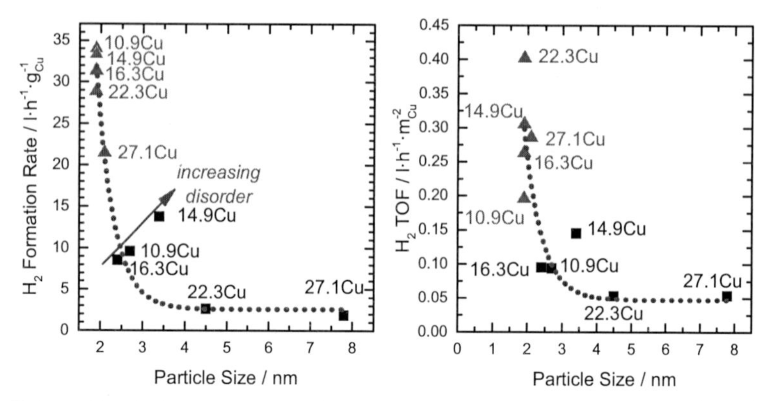

Figure II-20 Relationship between H_2 formation rates and average Cu metal particle sizes (left). Relationship between H_2 TOFs and Cu metal particle sizes (right). Cu metal particles sizes were obtained from XRD peak profile refinement. *Thin layer catalysts* are depicted as green triangles, and *thick layer catalysts* are depicted as black squares.

The fraction of surface atoms of a particle increases with decreasing particle size. This is illustrated in Figure II-21 showing various ideal cub octahedrons of different sizes. The contribution of edge atoms (dark blue balls), and their neighbors (grey balls) diminished with increasing particle diameter, while the amount of plane surface atoms (turquoise balls) grows rapidly with increasing particle size.

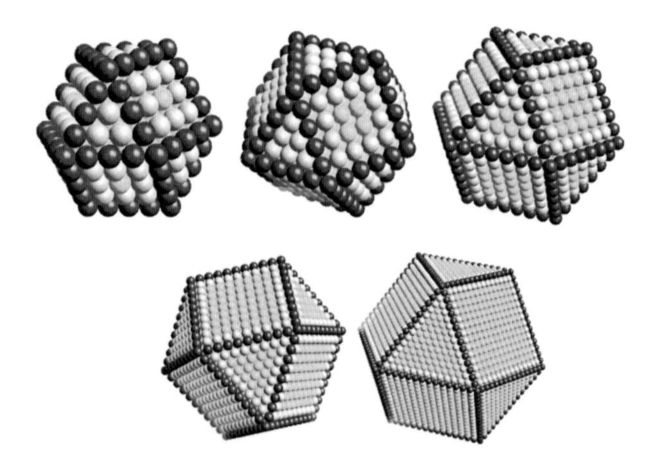

Figure II-21 Cu metal particles as cub octahedron with different types of Cu surface atoms. Diameter of particle is as follows: in line from left to right top: 2 nm (309 atoms), 2.6 nm (561 atoms), 3.6 nm (1415 atoms), and bottom: 5.1 nm (3871 atoms), and 7.6 nm (12431 atoms). Dark blue balls represents edge atoms, grey balls represents neighbor atoms of edge atoms, and turquoise balls represents surface atoms in plane.

The surface atoms are characterized by a reduced coordination spheres. Moreover, the surfaces of particles consist of different types of atoms such as atoms in plane faces, atoms at edges, and atoms at corners of a particle (compare Figure II-21). In real catalysts, defects yield additionally holes, terraces, and kinks reducing thus the fraction of atoms in plane faces. These unsaturated surface atoms exhibit different electronic structures, which may also influence the neighboring atoms.[100] Hence, the electronic structure of metal particles is determined by the fraction of surface atom types and therefore by the size and disorder of metal particle. Consequently, for Pt particles supported on TiO_2 non-metallic behavior was found for particle being smaller than 4.5 nm.[146] The electronic structure of catalysts strongly affects their catalytic activity. For instance, catalytic activity of Pt/SBA-15 catalysts in hydrogenolysis of ethane increased with decreasing particle size over a range from 1.8 to 7 nm.[147] Similar correlations between increasing catalytic activity and decreasing particle size were found for Co particles supported on silica employed in ethanol steam reforming.[103] The various Cu particle sizes found in Cu/SBA-15 catalysts investigated here could be responsible for the different H_2 formation rates and H_2 TOFs. Interestingly, H_2 TOFs correlated with H_2 formation rates (Figure II-22). The increased H_2 formation rates may therefore be attributed to increased intrinsic activities of Cu metal particles.

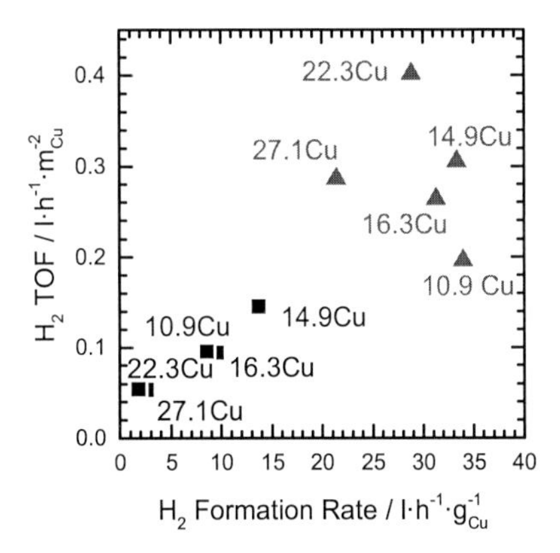

Figure II-22 Relationship between H_2 TOFs and H_2 formation rates of various Cu/SBA-15 catalysts. *Thin layer catalysts* are depicted as triangles and *thick layer catalysts* are depicted as squares. Corresponding Cu loading is given as number before *Cu*.

Additionally, the microstrain observed in small Cu metal particles also affected the electronic structure of the particles leading to increased activity.[99,143] Microstrain and static disorder may similarly affect the catalytic activity. The enhanced activity might thus be caused by increased static disorder. The correlation between H_2 formation rates or H_2 TOFs and EXAFS based disorder parameter is depicted in Figure II-23. The indicated defects may yield stacking faults evoking defects on the Cu metal particle surface.[6] They may also affect the electronic structure leading to enhanced catalytic activity.

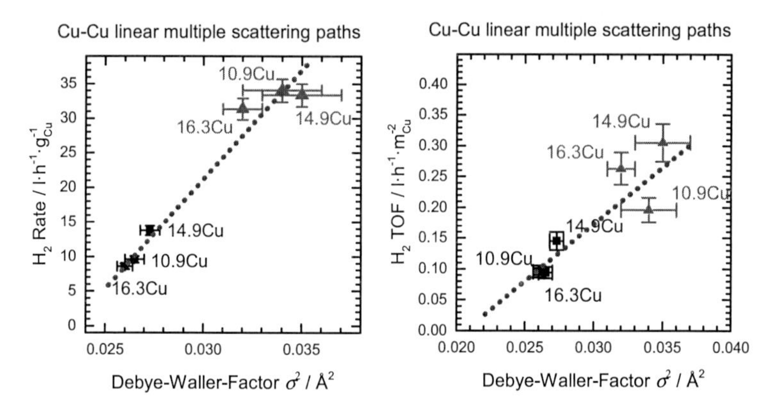

Figure II-23 H_2 formation rates and H_2 TOFs in dependence on DWFs of linear Cu-Cu multiple scattering paths, left and right, respectively. Triangles represent *thin layer catalysts* and squares represent *thick layer catalysts*. Error crosses assume 10 % variance of catalytic activities and 95 % certainty of DWFs.

The microstrain of Cu metal particles present in Cu/SBA-15 catalysts was apparently size related. It is therefore difficult to attribute the increased H_2 TOFs to either particle size and type of surface atoms or microstrain in particles. The data discussed in this chapter suggested a more complex relationship between structural features of supported Cu metal particles and for instance, interface with support and activity in methanol steam reforming.

II.3.15 Mechanism of methanol steam reforming

Cu/SBA-15 catalysts were employed in methanol steam reforming under optimized conditions. These are low contact times[139], conversion rates below 15 %[140], and temperatures of 250 °C[13] to suppress sintering of Cu metal particles[148]. Neglecting the proposed optimized reaction parameters resulted in increased conversion rates and CO production (data not shown). Following reaction parameters additionally affect the H_2 formation rates and were therefore considered in catalysts testing. The ratio between methanol and steam was kept constant at 1 at low conversion rates ensuring the surplus of methanol[149] and water[150]. The applied reaction conditions suppressed the reversed water gas shift reaction.[140] Decreasing the water/methanol ratio and tolerating reversed water gas shift reaction would lead to CO formation and thus to impurity of obtained H_2. Additionally, use of highly purified chemicals led to the needed oxygen free reaction feed[139] and prevented oxygen from participating in the reaction pathway.

Formaldehyde and methyl formate were found as reaction intermediates (compare Figure 0-7) similar to that Takahashi *et al.*[13] reported for Cu/SiO_2 catalysts working under comparable reaction conditions. The reaction pathway seemed to follow the methyl formate route proposed.[5,149] Formation of dioxomethylene as intermediate of a competing reaction pathway was not found. These findings were in good agreement with catalytic activities of other investigations of Cu/SiO_2 catalysts.[13,149] However, in these investigations methyl formate has not been found as adsorbed species on Cu/SiO_2 catalysts.[149] The initial step was apparently the dehydrogenation of methanol to methoxy bound on the Cu surfaces and released as formaldehyde. Formaldehyde was first detected and methyl formate was subsequently detected. According to Frank *et al.*[149] methyl formate is formed only in an excess of methanol (water/methanol ratio of 1 was sufficient). Formation of methyl formate is the rate determining step in methanol steam reforming over Cu/SiO_2 and $Cu/ZnO/Al_2O_3$ catalysts.[149] The here investigated Cu/SBA-15 catalysts showed catalytic activities being in good agreement to the proposed methyl formate route.

II.3.16 Deactivation of Cu/SBA-15 catalysts during methanol steam reforming

Cu/SBA-15 catalysts strongly deactivated within the first 12 h time on MSR stream. The deactivation of *thick layer catalysts* was more pronounced than that of *thin layer catalysts*. The strong deactivation of Cu based catalysts in methanol chemistry is well known. Cu metal particles on support materials, for example ZrO_2[141], Cu metal particles promoted by ZnO[5] or the combination of promoter and support in Cu/ZnO/SBA-15[53] catalyst exhibited strong deactivation in methanol steam reforming. Deactivation was attributed to sintering of Cu metal particles or coking resulting in loss of Cu surface area.[151] However, Cu catalysts typically show no coking. They are not prone to break C-O bonds and are not active in Fischer Tropsch reaction.[148] Moreover, reaction feed of MSR consisted partly of oxidizing compounds as water and freshly formed CO_2. Hence, deposition of freshly reduced carbon on Cu metal particles during MSR was not assumed. The low Hüttig temperature of Cu is given as possible reason for the deactivation behavior. The Hüttig temperature describes the onset of surface mobility of metal atoms and is about 177 °C.[152] According to Tammann[153], migration of Cu metal particles begins at temperatures higher than 405 °C and is further reduced for supported Cu metal particles[154]. Therefore it was advised to employ Cu catalysts below 300 °C.[148] XRD pattern of here investigated Cu/SBA-15 catalysts indicated invariant particle sizes after

6 h time on stream. Additionally, the hexagonal mesopore structure of SBA-15 limits Cu atoms and Cu metal particles migrating between mesopores. Agglomeration or sintering of Cu metal particles supported on SBA-15 was not observed during MSR at 250 °C.

Moreover, not only the increasing Cu loading tends to increased sintering[151] but also the distribution of particles on the SBA-15 influences sintering[53]. Well dispersed Cu/ZnO particles on support remain more catalytically active than particles, which are located next to other particles. Hence, Cu metal particles of here investigated *thin layer* and *thick layer catalysts* could be differently spread over the available surface areas of SBA-15. *Thin layer catalysts* seemed to exhibit well dispersed Cu metal particles in micropores and adjoining mesopores. The Cu metal particles might show in average longer distances to neighboring Cu metal particles in the same mesopores. The pore walls of SBA-15 apparently stabilized Cu metal particles. During the transformation of the well dispersed CuO_x particles of *thin layer precursors* into Cu metal particles in *thin layer catalysts* the extent of dispersion was preserved. *Thick layer catalysts* seemed to consist of Cu metal particles, which were closer to neighboring Cu metal particles. Comparing TEM pictures of freshly activated Cu/SBA-15 catalysts and TEM pictures of the same Cu/SBA-15 catalysts after reactivity testing may reveal the sintering depending on Cu metal particle distribution on SBA-15 surface as it was reported for Cu/ZnO/SBA-15 catalysts.[53]

Moreover, Campbell *et al.*[155] reported particle size effects, which should be considered for better understanding sintering processes of nanoparticles. They showed for Pb nanoparticles being smaller than 4 nm deviation in surface energies. Thus, for particles being smaller or larger than a threshold particle size different sintering processes were assumed although the chemical composition was identical. The different particle sizes of *thin layer* and *thick layer catalysts* (2 nm and 2.5 to 7 nm, respectively) may explain the different deactivation of *thin layer* and *thick layer catalysts*. Especially highly loaded 27.1Cu catalysts exhibited enlarged Cu metal particles and showed distinctly enhanced deactivation after 12 h time on stream.

Synthesis modalities were mainly decisive for deactivation of Cu/SBA-15 catalysts in MSR. Apparently, Cu metal particles of *thick layer catalysts* being poor in defects deactivated more rapidly than Cu metal particles of *thin layer catalysts* showing local disorder and microstrain. The Cu loading showed little influence on deactivation, but became more important at Cu loadings higher than 23 wt.%.

II.4 Conclusions

Cu/SBA-15 catalysts were model catalysts to study structure activity correlations between Cu metal particles and activity in methanol steam reforming. Varied calcination of the same precursor provided completely different Cu metal particles supported on SBA-15 making it a promising method for investigations of structure activity correlation of supported metal catalysts.

CuO_x/SBA-15 oxidic precursors were activated in 5 % H_2 at a temperature of 250 °C. The resulting Cu/SBA-15 catalysts were active in methanol steam reforming. Formation of intermediates such as formaldehyde and methyl formate indicated a reaction pathway according to the methyl formate route. The various structural characteristics of *thin* and *thick layer precursors* were preserved during activation leading to two types of Cu/SBA-15 catalysts exhibiting the same chemical composition (*chemical memory effect*). The first group comprised *thin layer catalysts* obtained from well dispersed, amorphous CuO_x particles of *thin layer precursors*. They possessed well dispersed and disordered Cu metal particles on SBA-15. The second group comprised thick layer catalysts obtained from larger CuOx particles of thick layer precursors. They possessed larger and less disordered Cu metal particles supported on SBA-15. The *thin layer catalysts* consisting of smaller and disordered Cu metal particles revealed higher H_2 formation rates and higher H_2 TOFs in methanol steam reforming compared to corresponding *thick layer catalysts*.

While *thin layer catalysts* indicated size-related disorder in Cu metal particles, *thick layer catalysts* showed no clear correlation between disorder in Cu metal particles and size of Cu metal particles. Deviation between Cu metal particle sizes derived from XRD and Cu surface area measurements indicated a large interface between the highly active disordered Cu metal particles of *thin layer catalysts* and the support. These strained Cu metal particles were probably located in the micropores of SBA-15. This was supported by less deactivation of *thin layer catalysts*. In contrast, Cu metal particles of *thick layer catalysts* were located mainly in the mesopores exhibiting smaller Cu/SiO_2 interfaces. This correlated with stronger deactivation of *thick layer catalysts* compared to corresponding *thin layer catalysts*.

Besides the disorder, the Cu metal particle size also correlated with the catalytic activity. Cu metal particles being smaller than 3 nm showed increased H_2 formation rates and increased H_2 TOFs. The increased H_2 formation rates correlated with increased H_2 TOFs. Therefore, increasing activity in methanol steam reforming was caused by both, increasing Cu surface areas and enhanced H_2 TOFs. Not only the dispersion but also the constitution of Cu metal particles played an important role in methanol steam reforming. Different reducibilities, different Cu surface areas of Cu metal particles of the same size in *thin layer catalysts* and different catalytic activities indicated modification of the Cu metal particle surfaces. Surface might vary with regard to the particle faces depending on various Cu metal particle shapes. Moreover, surfaces of Cu metal particles might possess unsaturated Cu atoms or different amounts of defects. The entangling of Cu metal particle size, disorder and possible variation of Cu particle shape apparently affected the electronic structure of Cu metal particles. Hence, various activities in methanol steam reforming were observed for various Cu/SBA-15 catalysts.

The results presented here gave rise for further investigations especially with focus on the role of ZnO in commercially $Cu/ZnO/Al_2O_3$ catalysts. Therefore, new model catalysts are required consisting of both, ZnO and Cu metal particles exhibiting sizes reported here.

Chapter III Impact of redox pretreatment on structure and activity of Cu/SBA-15 catalysts in methanol steam reforming

III.1 Introduction

Methanol steam reforming (MSR) releases H_2 from methanol. MSR proceeds well over Cu based catalysts at ambient pressure and 250 °C, while H_2 production can be improved by temporary addition of O_2 to the MSR feed.[11,23,25,156] In the previous, it was demonstrated that catalytic activity of Cu/SBA-15 catalysts correlated with the structure of Cu metal particles. Performance in methanol steam reforming depended strongly on Cu metal particle size and microstrain in Cu metal particles. However, Cu/SBA-15 catalysts deactivated as it has been reported for other Cu catalysts.[5,10,53] Regeneration of catalysts is often achieved by oxidative treatment.[157] For instance, supported Co metal particles can be re-dispersed employing oxidative regeneration yielding increased activity.[158] Active sites of catalysts are regained due to cleaning, or due to re-dispersion of metal particles.[158]

The activity of frequently used $Cu/ZnO/Al_2O_3$ catalyst can be regained by oxidative regeneration.[156] Interestingly, temporary addition of oxygen to methanol steam reforming feed also led to increased H_2 formation.[11,25,156] The enhanced H_2 production occasionally exceeded the level of initial H_2 production. However, after repeating oxygen co-feeding for several times no further enhanced activity was observed anymore.[141] Structural changes in Cu metal particles after oxygen co-feeding depended on the support material used. For example, Cu metal particles of Cu/ZnO catalysts exhibited increased microstrain, after oxygen co-feeding. Although Cu metal particle sizes increased after oxygen co-feeding enhanced catalytic activity was observed.[11] After oxidative regeneration of $Cu/ZnO/Al_2O_3$ sintering of Cu metal particles was not detected. Catalysts regained their catalytic activity and deactivated after each regeneration cycle.[156]

Here, Cu/SBA-15 model catalysts were investigated to determine the influences of redox treatment on structure of silica supported Cu metal particles. Starting from the *thin* and *thick layer precursors*, investigations focused on the changes in structure evoked by oxygen co-feeding during methanol steam reforming. Furthermore, a redox cycle was applied to Cu/SBA-15 catalysts before catalytic testing, to compare the specific Cu surface areas as result of different activation procedures. The structure of Cu metal particles was also investigated as well as the dispersion of Cu metal particles with regard to activation procedure.

III.2 Experimental

III.2.1 In situ X-ray diffraction (XRD) during methanol steam reforming and temporary oxygen co-feeding

Samples were measured at the setup described in *II.2.3*. The oxidic precursors were activated by temperature-programmed reduction in a flow of 100 ml/min of 5 % H_2 balanced by He. Samples were heated to 250 °C at 5 °C/min. Methanol steam reforming over activated Cu/SBA-15 catalysts was tested at 250 °C in a flow of 100 ml/min feed consisting of 2 % MeOH and 2 % H_2O balanced by He. A stream of 5 ml/min of pure O_2 was added to the reaction feed for 5 min (Figure III-1). XRD patterns were recorded during activation, after activation, before O_2 co-feeding, twice after O_2 co-feeding, as well as after cooling. Measurement parameters of XRD scans are given in 51*II.2.3*.

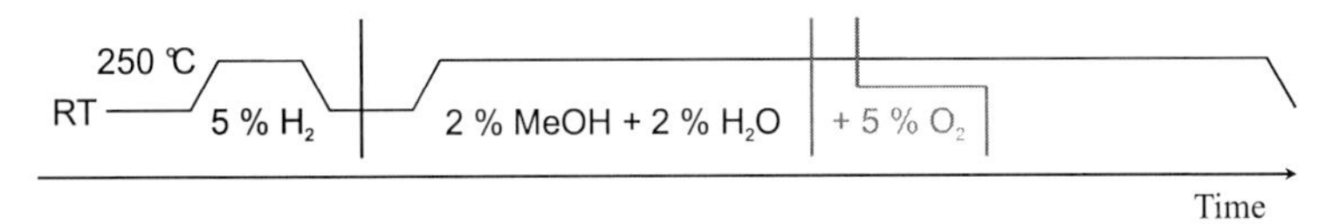

Figure III-1 Standard activation route for Cu/SBA-15 catalysts starting from oxidic precursors. Temporary addition of O_2 to methanol steam reforming feed is illustrated.

III.2.2 In situ X-ray absorption (XAS) during methanol steam reforming and temporary O_2 addition

EXAFS spectra were measured at beamline X at HASYLAB, Hamburg. During heating and oxygen co-feeding the QEXAFS mode was used. Detailed parameters and sample preparation are given in *II.2.4*. Cu/SBA-15 catalysts were activated by reducing oxidic precursors in a flow of 30 ml/min 5 % H_2 balanced by He. Therefore, catalysts were heated to 250 °C at 5 °C/min and held for 45 min at 250 °C. Freshly activated catalysts were tested in methanol steam reforming. Reaction feed of 2 % MeOH and 2 % H_2O balanced by He flew with 30 ml/min over Cu/SBA-15 catalysts. At 250 °C, a flow of 10 ml/min of 20 % O_2 balanced by He was added to reaction feed for approximately 5 min (Figure III-2).

III.2.3 Redox activation of Cu/SBA-15 catalysts and Cu surface areas and activity in methanol steam reforming

Redox activation consisted of a prior redox cycle applied to oxidic precursors and the standard activation. The processes of standard activation and redox activation are compared in Figure III-2 on the top in black and on the bottom in red, respectively. Oxidic precursors were reduced in a flow of 40 ml/min of 5 % H_2 balanced by He during heating at 5 °C/min up to 250 °C and held for 2 h at 250 °C. After cooling, activated Cu/SBA-15 samples were re-oxidized in a flow of 40 ml/min of 5 % O_2 balanced by He. 250 °C were reached at 5 °C/min and samples were held for 30 min at 250 °C. Subsequently, the activation step was repeated. Freshly activated catalysts were examined with regard to Cu surface area and catalytic activity in methanol steam reforming. Cu surface areas were determined following the procedure described in *II.2.1*. Testing catalysts in methanol steam reforming was performed according to instruction given in *II.2.2*.

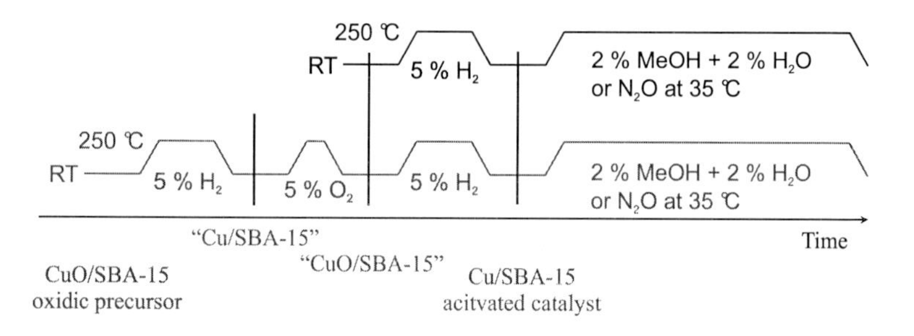

Figure III-2 Standard activation of SBA-15 catalysts in black on the top and the redox activation in red on the bottom. After activation either methanol steam reforming at 250 °C or determination of Cu surface areas followed.

III.2.4 In situ X-ray absorption spectroscopy (XAS) at the Cu K edge during redox activation

Cu/SBA-15 samples were prepared as described in II.2.4. Samples were measured at the Cu K edge at beamline C at HASYLAB at DESY, Hamburg. The energy range of 8.900 keV to 9.940 keV was scanned using a Si(111) double crystal monochromator. During the heating ramps, QEXAFS measurements were performed in the range of 8.920 keV to 9.300 keV. During the QEXAFS scan, the monochromator moved continuously. QEXAFS scan time was about 35 s. Redox activation was performed similar to laboratory experiments. Heating ramps at 5 °C/min were used and gas flows of 30 ml/min were employed. Samples were held for approximately 40 min at elevated temperature. The experimental procedure is illustrated in Figure III-2. First, oxidic precursors were reduced at 250 °C in 5 % H_2 balanced by He. After cooling to room temperature, re-oxidation was performed in 5 % O_2 containing He atmosphere at 250 °C. In the following, the samples were once more activated at 250 °C in 5 % H_2 balanced by He. EXAFS spectra were measured before and after each ramp at elevated temperature and room temperature.

III.3 Results and discussion

III.3.1 Increased activity after temporary O_2 addition to methanol steam reforming feed

After temporary co-feeding of oxygen during methanol steam reforming, increased H_2 production was measured for Cu/SBA-15 catalysts. In Figure III-3, the ion currents of H_2 are depicted as green lines for *thin layer catalysts* (left) and as black lines for *thick layer catalysts* (right). After reaching a temperature of 250 °C, H_2 ion currents decreased. During temporary O_2 addition, H_2 production was suppressed. After removal of O_2 from the feed, H_2 production increased to a level higher than the initial level for 10.9Cu *thin layer* and *thick layer catalysts*, and 16.3Cu *thick layer catalyst*. 14.9Cu and 16.3Cu *thin layer catalysts* reached the initial level of H_2 production after employing oxygen co-feeding. The 16.3Cu *thick layer catalyst* did not reach the initial level of H_2 formation but also showed increased H_2 production after co-feeding of oxygen.

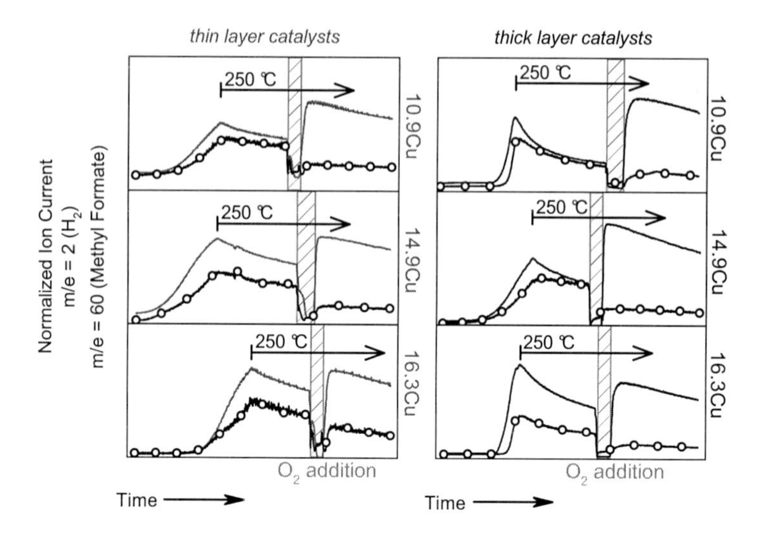

Figure III-3 Evolution of H_2 production (green/black line) and methyl formate production (brown line with empty circles) during methanol steam reforming in 2 % MeOH and 2 % H_2O over *thin layer* (left) and *thick layer catalysts* (right) in *in situ* XAS cell. Starting point of isothermal period at 250 °C is illustrated by arrows. Temporary O_2 addition is marked as striped block.

Interestingly, the ion current of the sideproduct methyl formate (brown line with empty circles) decreased after oxygen co-feeding. Despite similar deactivation of methyl formate and H_2 before oxygen co-feeding, methyl formate was produced less after oxygen co-feeding, while H_2 production increased. Hence, O_2 addition led not only to increased number of active sites but also to other product distributions, indicating changes in the reaction pathway.

III.3.2 Evolution of Cu metal particles during temporary O_2 addition observed by in situ XAS

Time resolved QEXAFS scans at the Cu K edge revealed the transformation of Cu metal particles present on SBA-15 during temporary O_2 addition to methanol steam reforming feed. The normalized XANES spectra of 14.9Cu *thin layer* and *thick layer catalysts* are depicted in Figure III-4 and Figure III-5, respectively. Corresponding plots of 10.9Cu and 16.3Cu catalysts are shown in Figure 0-8 to Figure 0-11 in *Appendix*.

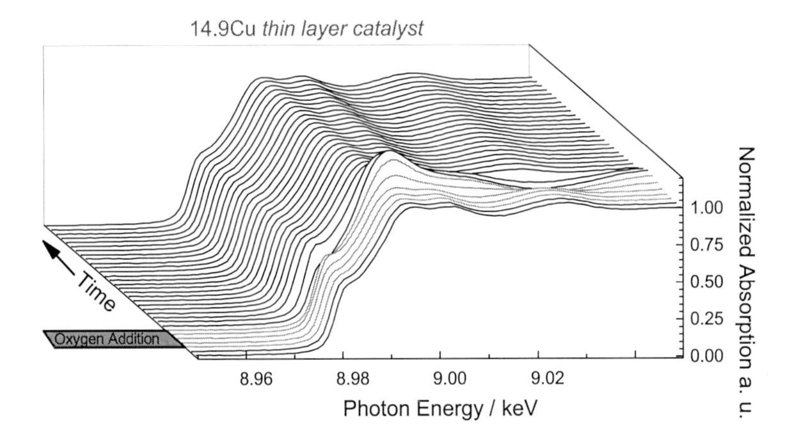

Figure III-4 Evolution of XANES spectra of 14.9Cu *thin layer catalyst* at Cu K edge during methanol steam reforming at 250 °C in a stream of 30 ml/min of 2 % MeOH and 2 % H_2O balanced by He, and employed temporary O_2 addition of 5 % in a total stream of 40 ml/min. Temporary O_2 addition is marked as orange area and orange spectra.

During methanol steam reforming the double peak in the range of 8.990 keV to 9.000 keV indicated the presence of Cu metal particles present on SBA-15. Shortly after addition of O_2 to the feed, XANES spectra at Cu K edge changed. Changes occurred independently of Cu loading but depended on calcination mode. Both, Cu metal particles of *thin layer* and *thick layer catalysts* were oxidized. The double peak in XANES spectra was changed into a single peak, white line (orange colored spectra). After removal of O_2 from the methanol steam reforming feed, re-reduction started immediately. The white line rapidly decreased and finally transformed into the double peak indicating Cu metal particles present on SBA-15 after temporary O_2 addition to methanol steam reforming feed.

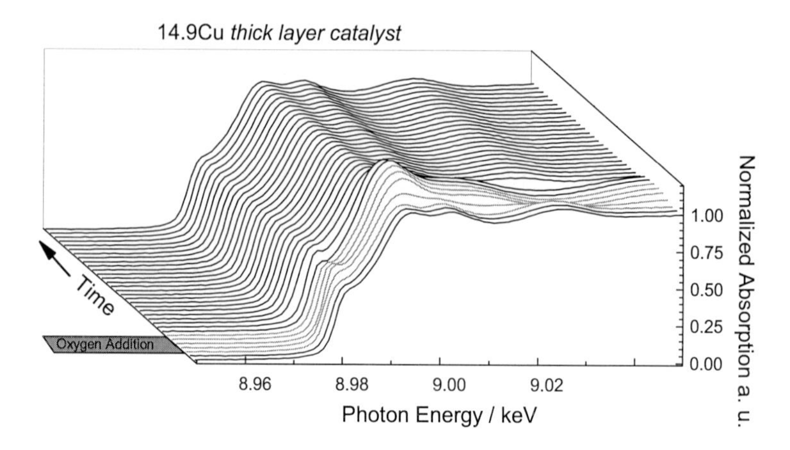

Figure III-5 Evolution of XANES spectra of 14.9Cu *thick layer catalyst* at Cu K edge during methanol steam reforming at 250 °C in a stream of 30 ml/min of 2 % MeOH and 2 % H_2O balanced by He and employed temporary O_2 addition of 5 % in a total stream of 40 ml/min. Temporary O_2 addition is marked as orange area and orange spectra.

The evolution of Cu metal particles of the differently calcined 14.9Cu catalysts showed differently shifted absorption edges in XANES spectra due to oxygen in feed. The degree of oxidation varied under the same reaction conditions depending on the calcination mode. This can be seen in Figure III-6. The phase contributions of reference compounds are depicted, which were obtained from linear combination XANES fits described in *II.3.2*.

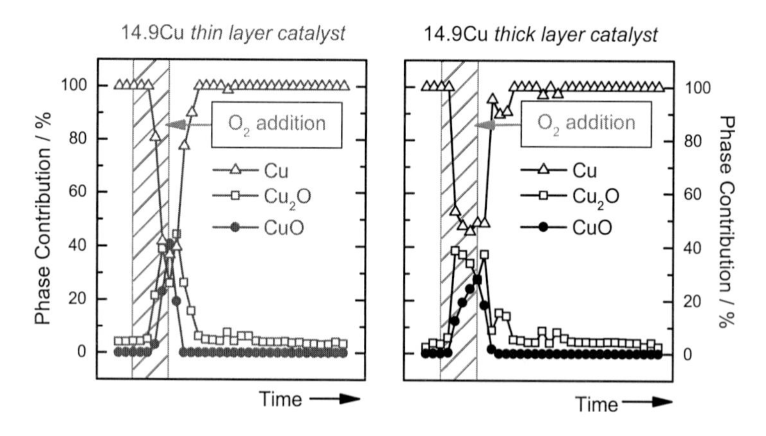

Figure III-6 Results of linear combination XANES fit using reference XANES spectra of Cu foil (triangle), Cu_2O (squares), and CuO (circles) in the range of 9.895 keV to 9.000 keV. Period of O_2 addition is drawn as striped column.

The 14.9Cu *thin layer catalyst* showed less remaining Cu (25 %) than 14.9Cu *thick layer catalysts* (45 %). Simultaneously, the maximum of formed CuO (40 %) was higher in 14.9Cu *thin layer catalyst* than in 14.9Cu *thick layer catalyst* (29 %). The contribution of Cu_2O (of about 40 %) was similar in both catalysts. The differently calcined 10.9Cu and 16.3Cu catalysts exhibited the same trends in phase fractions during the temporary addition of O_2 to the feed. Therefore, the Cu metal particles of all *thin layer catalysts* showed an increased reactivity. The apparently slower oxidation of Cu metal particles in *thick layer catalysts* can be attributed to enlarged Cu metal particles present in *thick layer catalysts*. The XANES spectra of catalysts before and after oxygen co-feeding were identical (Figure III-4 and Figure III-5) indicating no distinct changes in Cu metal particles.

III.3.3 Impact of oxygen co-feeding on Cu metal particles observed by in situ XRD

The partial oxidation of supported Cu metal particles during oxygen addition to MSR feed and the subsequent re-reduction after oxygen removal resulted in Cu metal particles. These re-reduced Cu metal particles exhibited little changes in their constitution compared to initial Cu metal particles. Profile analysis of XRD pattern was used to reveal differences in the structure of Cu metal particles. Therefore, the analysis of the Cu(111) and Cu(002) diffraction peaks introduced in *II.3.6* was performed for all Cu/SBA-15 catalysts during activation of the oxidic precursors, during MSR, and during MSR after temporary O_2 addition to the MSR feed. The results are summarized in Figure III-7.

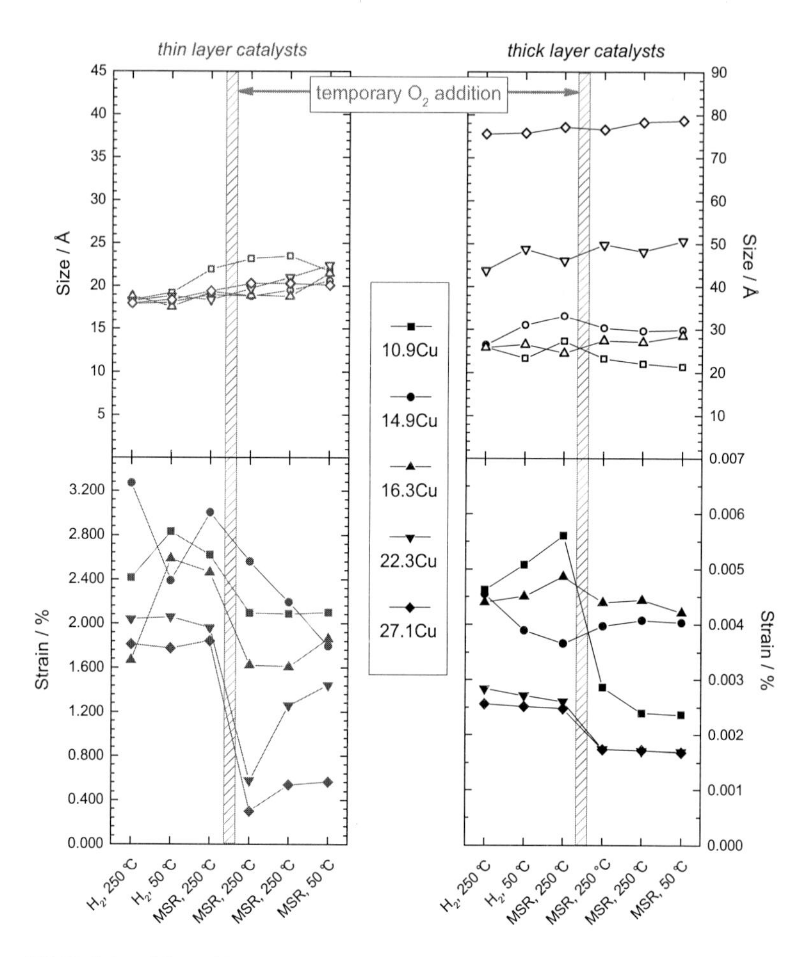

Figure III-7 Evolution of Cu particle size (top) and microstrain (bottom) obtained from Cu(111) XRD peak profile analysis of *thin layer* and *thick layer catalysts*, left and right, respectively. Information is given for Cu metal particles during activation (H_2) and methanol steam reforming (MSR), before and after temporary O_2 addition to methanol steam reforming feed (striped column).

The Cu particle sizes are given on the top left and right for *thin layer* and *thick layer catalysts*. The size of Cu metal particles was nearly invariant. A little increase of Cu particle sizes was slightly indicated for *thin layer catalysts*. However, the particle growth was neither distinct nor correlated with O_2 addition to the MSR feed (striped columns). The microstrain of Cu metal particles indicated systematic changes. For all *thin layer catalysts*, the microstrain of re-reduced Cu metal particles tended to decrease after oxygen co-feeding. Interestingly, 27.1Cu, 22.3Cu, and 16.3Cu *thin layer catalysts* showed re-increasing microstrain after longer period of operation (Figure III-7, bottom left). However, the initial level of microstrain was not reached. Even the less strained Cu metal particles of 10.9Cu, 22.3Cu, and 27.1Cu *thick layer catalysts* exhibited further decrease of microstrain after oxygen co-feeding. The microstrain of 14.9Cu and 16.3Cu *thick layer catalyst* was apparently unaffected (Figure III-7, bottom right). In contrast to *thin layer catalysts*, an increase of microstrain in *thick layer catalysts* after 12 h time on stream was not observed.

III.3.4 Effect of oxygen co-feeding on Cu/SBA-15 catalysts

The oxygen co-feed as surplus (5 % O_2 to 2 % H_2O and 2 % MeOH balanced by He) led to deactivation of Cu/SBA-15 catalysts due to oxidation of Cu metal particles. Removing the oxygen from the MSR feed resulted in rapid re-formation of Cu metal particles exhibiting increased H_2 formation rates. Reduced Cu particle sizes after co-feeding of oxygen were not found. Hence, no re-dispersion of Cu metal particles was observed, which would explain the increased H_2 formation rates due to increased Cu surface area. Moreover, after co-feeding of oxygen less microstrain in Cu metal particles was indicated. Simultaneously, increased H_2 production was measured. This was contradictory to previously observed relations (compare *Chapter II*). Therefore, the oxygen co-feed was simulated as redox pretreatment before methanol steam reforming. The redox pretreatment was performed prior to methanol steam reforming described in *III.2.3* to elucidate structural changes in Cu metal particles, being once or twice activated.

III.3.5 Temperature-programmed reduction (TPR) during redox activation

According to the presented activation modes (compare *III.2*), the Cu metal particles were either once reduced or twice reduced and in-between re-oxidized to simulate the temporary O_2 co-feeding. Figure III-8 illustrates the H_2 consumption during different activation steps. The first reduction of *thin layer precursors* is represented as green line (left). The first reduction of *thick layer precursors* is represented as black line (right). The corresponding second reduction during redox activation is depicted as red line on both sites of Figure III-8. After first reduction and subsequent re-oxidation, the H_2 consumption peak of second reduction (red lines) shifted to lower temperatures. The extent of the shift was similar for all *thin layer precursors*. In contrast, *thick layer precursors* showed various shifts, especially 22.3Cu and 27.1Cu *thick layer precursors* exhibited more pronounced shifts of the H_2 consumption peaks.

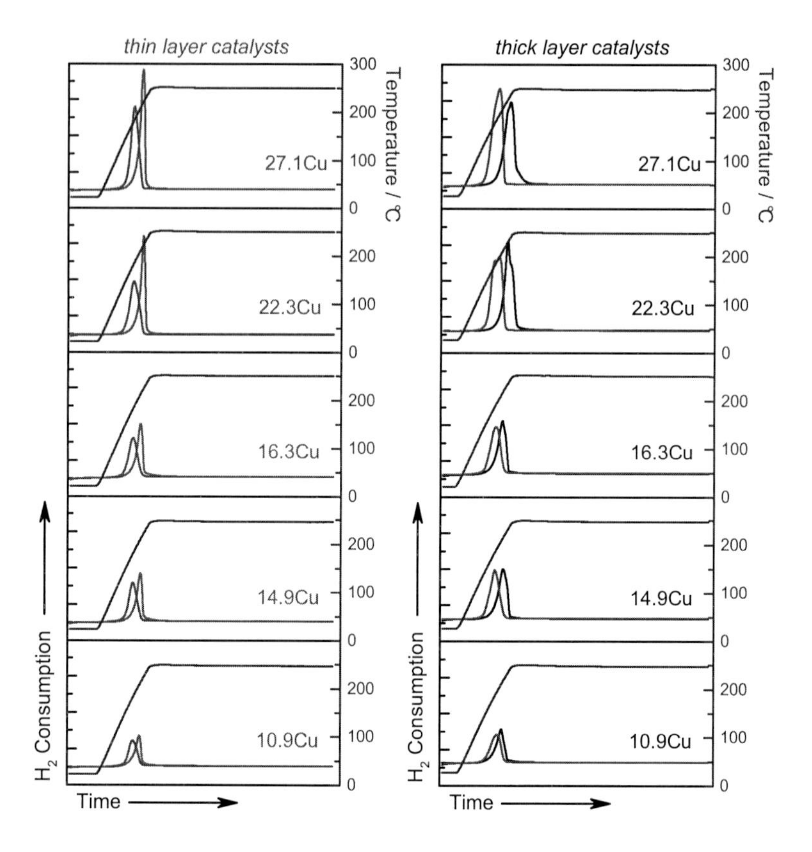

Figure III-8 H_2 consumption during redox activation of *thin layer* and *thick layer catalysts* left and right, respectively. H_2 consumption peaks of first reduction of oxidic precursors are plotted in green and black lines. H_2 consumption peaks of second reduction are depicted in red.

Moreover, the H_2 consumption peak profiles measured during the second reduction of *thin layer precursors* differed from those of the first reduction. The H_2 consumption peaks of the second reduction showed an increased symmetry. The first derivatives of H_2 consumption peaks measured for the 22.3Cu *thin layer catalyst* are depicted in Figure III-9 (left). In Figure III-9 (right) first derivatives of the H_2 consumption peaks of the 22.3Cu *thick layer catalyst* are depicted. The increased number of local maxima of first derivatives of the *thick layer catalyst* compared to derivatives of the *thin layer catalyst* indicated two step reduction for *thick layer precursors* (compare *II.3.1*) The first derivative of H_2 consumption of *thick layer catalysts* during the second reduction differed also from that measured during the first reduction. Interestingly, the first derivatives of H_2 consumptions of the second reduction during redox activation of *thin layer* and *thick layer precursors* resembled each other more than the corresponding derivatives measured during first reduction. The redox cycle enhanced the reducibility of CuO_x particles supported on SBA-15. Hence, either the reduction process[17], the CuO_x particle size distribution[10,121], the disorder of CuO_x particles[126], or a combination of those was changed.

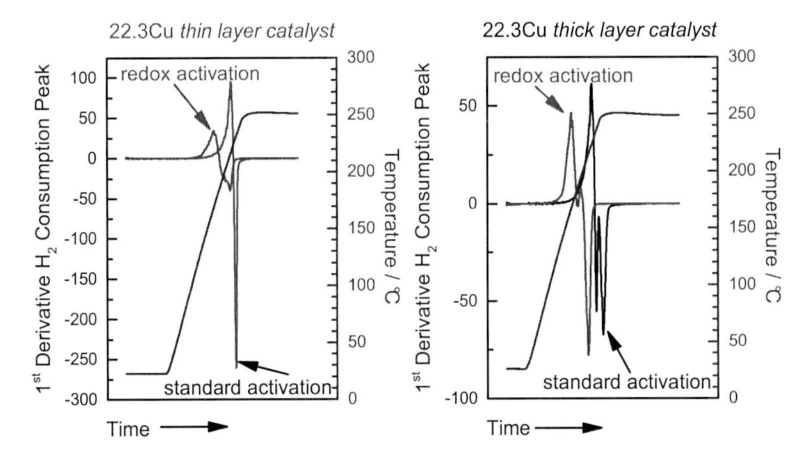

Figure III-9 Comparisons of the 1st derivatives of H_2 consumption during standard activation also denoted as first reduction (green, left; black, right) and the second reduction of redox activation (red) for 22.3Cu *thin layer* and *thick layer catalysts*, left and right, respectively.

The amount of H_2 needed during reduction was similar for each catalyst during first or second reduction. Hence, no additional reducible species were accessible on SBA-15 and the re-oxidation of Cu metal particles proceeded completely to Cu(II) ions in CuO_x particles. The observed changes may be related to the interplay of CuO_x particles and SBA-15 support or to structural changes of CuO_x particles.

III.3.6 Phase transformation during first and second reduction of redox activation using LC-XANES fit

The XANES spectra at the Cu K edge of Cu/SBA-15 catalysts measured during redox activation were fitted using reference XANES spectra of Cu foil, Cu_2O and CuO. In Figure III-10 results are given for the 14.9Cu *thin layer* and *thick layer catalysts*. The onset temperature of Cu_2O formation was shifted from 170 °C (filled squares) during first reduction to 145 °C (empty squares) during second reduction. The onset of further reduction to Cu was also shifted to lower temperatures. During the first reduction, Cu was found at 200 °C and temperatures above (filled triangles), while during the second reduction Cu was found at 170 °C and temperatures above (empty triangles).

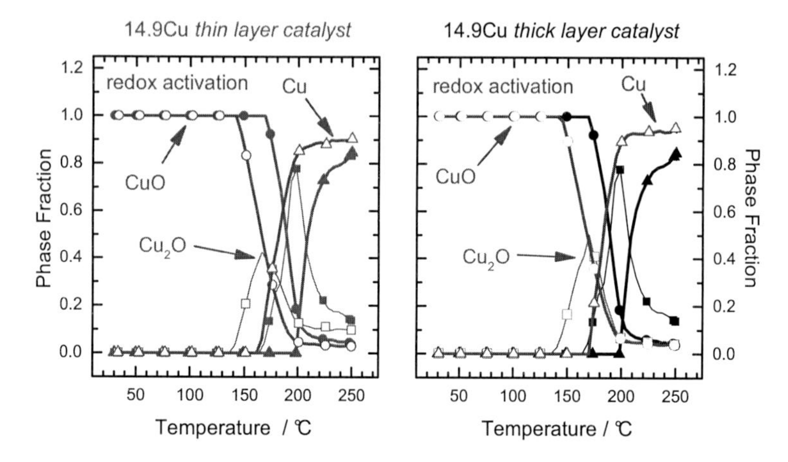

Figure III-10 Evolution of phase composition of 14.9Cu *thin layer* (left) and *thick layer catalyst* (right) during standard activation (green and black, filled symbols) and second reduction during redox activation (red, empty symbols). Arrows indicate traces of second reduction. Phase fraction of Cu (triangle), Cu_2O (squares), and CuO (circles) derived from LC-XANES fit in the range of 8.950 keV to 9.000 keV at the Cu K edge using corresponding reference spectra.

The maximum contribution of intermediate Cu_2O decreased from 0.8 during the first reduction to 0.5 during the second reduction. Moreover, the fraction of formed Cu phase increased from first to second reduction. The apparent phase fraction of Cu_2O decreased equally after second reduction. This revealed an improved reducibility of CuO_x particles supported on SBA-15 obtained after first redox cycle. The two step reduction was preserved. Thus the improved reducibility indicated distinct structural differences in the CuO_x particles of intermediate CuO_x/SBA-15 after the redox cycle compared to those of oxidic precursors.

III.3.7 XANES of oxidic precursors and intermediate stage CuO_x/SBA-15

The analysis of XANES spectra of oxidic precursors was discussed in detail in *I.3.8*. Accordingly, the first derivatives of normalized XANES spectra of oxidic precursors (green and black), and intermediate stage CuO_x/SBA-15 (red) are depicted in Figure III-11. XANES spectra of *thin layer* and *thick layer* samples are depicted left and right, respectively.

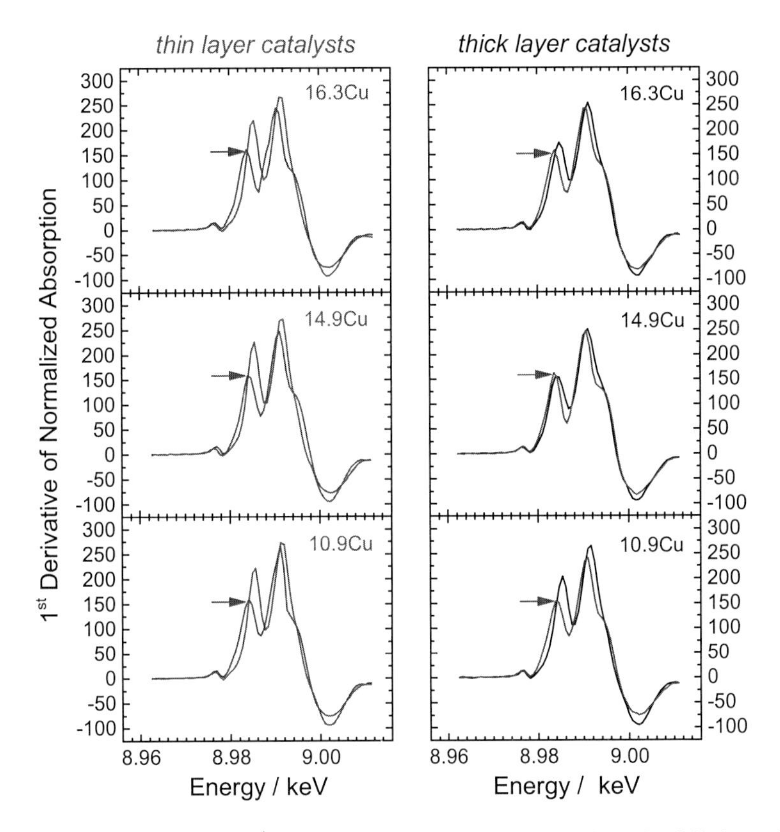

Figure III-11 Illustration of the 1st derivatives of normalized XANES spectra at Cu K edge of *thin layer* and *thick layer* CuO_x/SBA-15, left and right, respectively. Data obtained from oxidic precursors are drawn as green and black line. Corresponding data of intermediate CuO_x/SBA-15 are drawn as red line. Additionally, the second maxima of intermediate CuO_x/SBA-15 are indicated by arrows.

XANES spectra of CuO_x/SBA-15 before and after redox cycle showed the invariant 1s→3d transition at 8.977 keV. However, the 1s→4p transitions ranging from 8.981 keV to 8.991 keV were found at lower energies after the redox cycle. Especially, the 1s→$4p_z$ shake-down transition at about 8.983 keV shifted distinctly to lower energies after redox cycle. The corresponding maxima are indicated with arrows in Figure III-11. This shift was more pronounced in *thin layer* CuO_x/SBA-15 intermediate. Moreover, at 8.995 keV a shoulder was observed in CuO_x/SBA-25 intermediates. In contrast to *thin layer precursors*, *thick layer precursors* exhibited CuO like nanoparticles present on SBA-15 (compare *I.3.16*). After the redox cycle, XANES spectra of intermediate CuO_x/SBA-15 catalysts resembled that of CuO reference. Hence, the various CuO_x particles of oxidic precursors were transformed into CuO like nanoparticles after redox cycle.

III.3.8 EXAFS at Cu K edge of oxidic precursors and intermediate CuO_x/SBA-15

The local structure in CuO_x particles was investigated to understand the various reducibilities of oxidic precursors and intermediates CuO_x/SBA-15. The $FT(\chi(k)\cdot k^3)$ of *thin layer* and *thick layer precursors* are depicted in Figure III-12, left and right, respectively. Oxidic precursors are printed as dotted line in comparison to their corresponding $FT(\chi(k)\cdot k^3)$ of intermediate CuO_x/SBA-15 (red lines). $FT(\chi(k)\cdot k^3)$ of intermediate CuO_x/SBA-15 differed from those of corresponding oxidic precursors. The differences were more pronounced in *thin layer precursors* agreeing well with the extent of changes in XANES spectra.

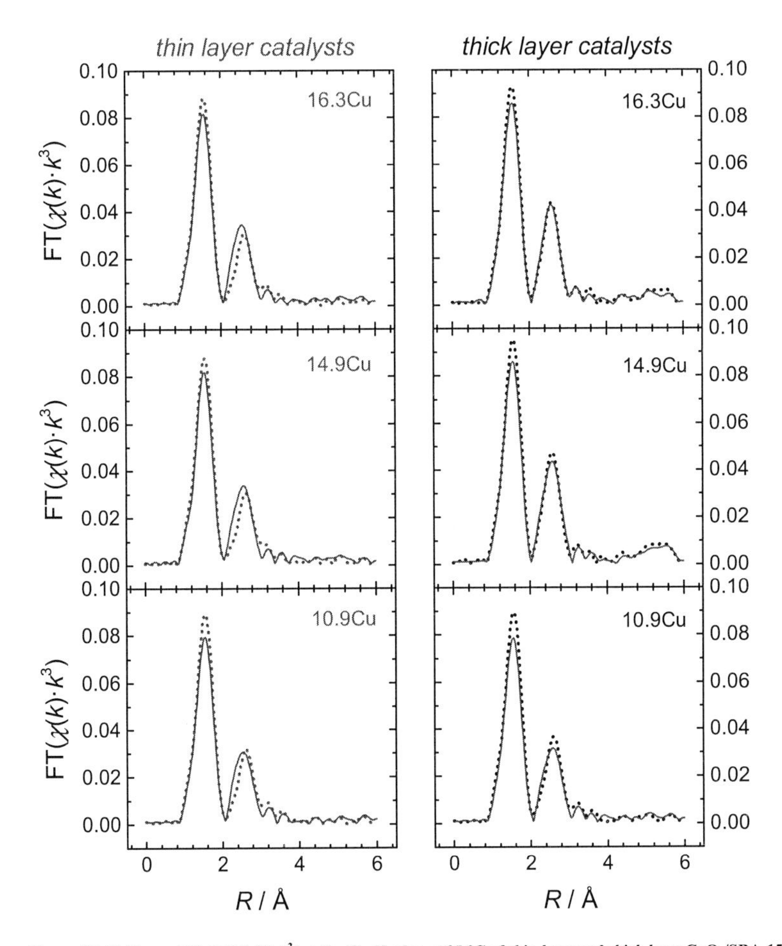

Figure III-12 Phase shifted FT($\chi(k)\cdot k^3$) at the Cu K edge at 35 °C of *thin layer* and *thick layer* CuO_x/SBA-15, left and right, respectively. Oxidic precursors are printed as dotted line and corresponding freshly re-oxidized intermediate CuO_x/SBA-15 after redox cycle are printed as red line. Fourier transformation was performed over the range of $3 < k < 14$ Å^{-1}.

In all FT($\chi(k)\cdot k^3$), the amplitude of the first peak was reduced after the redox cycle. Moreover, the amplitude of the second peak was slightly decreased in the FT($\chi(k)\cdot k^3$) of *thick layer precursors* after the redox cycle. In the FT($\chi(k)\cdot k^3$) of *thin layer precursors*, the second peak changed its shape and shifted to shorter distances R after the redox cycle. The reduced amplitude of the first peak corresponded to the atom distances between Cu and four oxygen atoms forming a square plane around the Cu atom. During reduction, the oxygen ions needed to be removed to form Cu metal particles. According to the fit procedure introduced in *I.3.11* for *thick layer catalysts*, the refined Debye-Waller-Factors (DWFs) of the first Cu-O scattering paths are depicted in Figure III-13. Since all spectra were measured at the same temperature, the DWF represented the static disorder in the local structure. The intermediate CuO_x/SBA-15 samples exhibited distinctly increased DWFs (Figure III-13, empty symbols) indicating a higher degree of disorder in [CuO_4] square plane. Interestingly, the differences of DWFs among oxidic precursors (Figure III-13, filled symbols) were preserved after redox cycle. Therefore, the processes during redox cycle apparently affected the structure of CuO_x particles equally in *thin layer* and *thick layer precursors*.

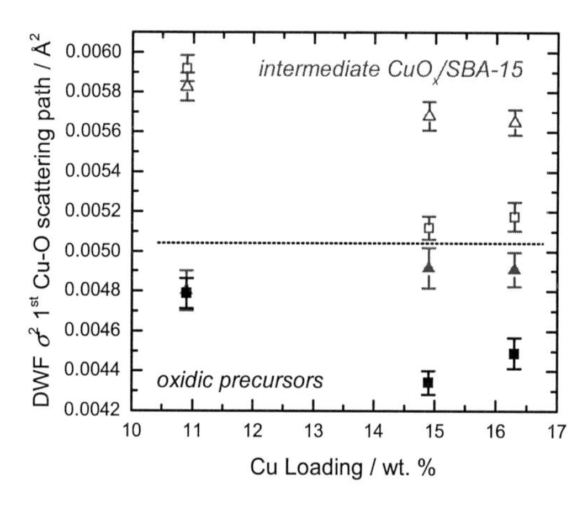

Figure III-13 Debye-Waller-Factors (DWFs) of first Cu-O scattering path of *thin layer precursors* (filled triangles) and *thick layer precursors* (filled squares) and the corresponding intermediates CuO_x/SBA-15 as empty symbols. Errors are given with 95 % certainty.

Reduction of *CN*s, which may also cause decreased amplitude, was unlikely. Cu^{2+} ions are prone to form a square plane (*CN* = 4) or distorted octahedron (*CN* = 4 + 2).[159] Therefore, the increased static disorder might base on distortion of the [CuO_4] square plane.

III.3.9 Correlation between static disorder in CuO_x particles and reducibility

Increase of disorder in Ni and Cu silicates resulted in increased reducibility.[122] Increased number of defects in CuO films led also to increased reducibility.[126] Similar behavior was found for CuO_x/SBA-15 samples investigated here. The correlation between the static disorder parameter (DWF) and the maximum in H_2 consumption during activation is illustrated in Figure III-14.

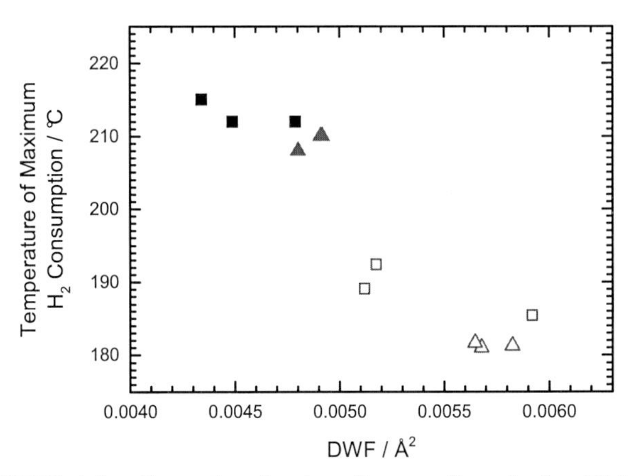

Figure III-14 Evolution of temperature of maximum H_2 consumption as function of Debye-Waller-Factor (DWF) of the first Cu-O backscattering path. *Thick layer* samples are depicted as squares and *thin layer* samples are depicted as triangles. Filled symbols represent oxidic precursors and empty symbols represent intermediate CuO_x/SBA-15 samples.

With increasing static disorder, the maximum of H_2 consumption peak was observed at lower temperatures. Accordingly, improved reducibility was mainly caused by increasing disorder in CuO_x particles. A significant reduced CuO_x particle size was therefore unlikely. This was supported by similar Cu metal particle sizes after both activations (see below). Hence, sintering of Cu metal particles due to redox cycle[11] was not observed on Cu/SBA-15 catalysts investigated here.

III.3.10 EXAFS of Cu/SBA-15 catalysts after standard activation and after redox activation

In Figure III-15, the $FT(\chi(k)\cdot k^3)$ of *thin* and *thick layer catalysts* measured after first reduction are depicted in green and black, respectively. The corresponding $FT(\chi(k)\cdot k^3)$ measured after the second reduction are colored red. $FT(\chi(k)\cdot k^3)$ of the two Cu/SBA-15 catalyst possessing the same Cu content superimposed except of little deviations at distances of 5 Å.

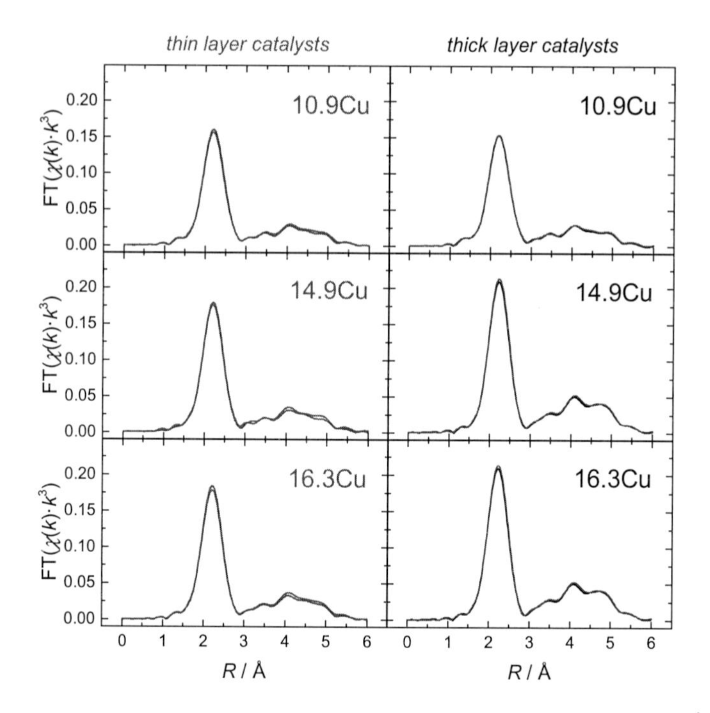

Figure III-15 FT($\chi(k)\cdot k^3$) of *thin layer catalysts* (left) and *thick layer catalysts* (right) at the Cu K edge after normal activation (green and black) and after redox activation (red) measured at 35 °C.

To compare the little deviations in the FT($\chi(k)\cdot k^3$) an EXAFS refinement was performed according to procedure described in *II.3.5.* The results are summarized in Figure III-16. Parameters of refined spectra of *thick layer catalysts* were invariant after first and after second reduction. This indicated that the structure of Cu metal particles of 10.9Cu, 14.9Cu, and 16.3Cu *thick layer catalysts* was not affected by the redox cycle.

Parameters of fitted spectra of *thin layer catalysts* indicated similar Cu metal particles before and after the redox cycle. Cu particle sizes were identical indicated by *CN*s and *R*s. But the results suggested that static disorder in Cu metal particles decreased after the second reduction compared to the static disorder found after the first reduction. Remarkably, a decreased static disorder was indicated by DWFs of multiple scattering paths. Especially the linear multiple scattering paths being sensitive towards disorder[160], exhibited diminished DWFs (Figure III-16, bottom left). The first Cu-Cu scattering path revealed no changes in the microstructure of Cu metal particles.

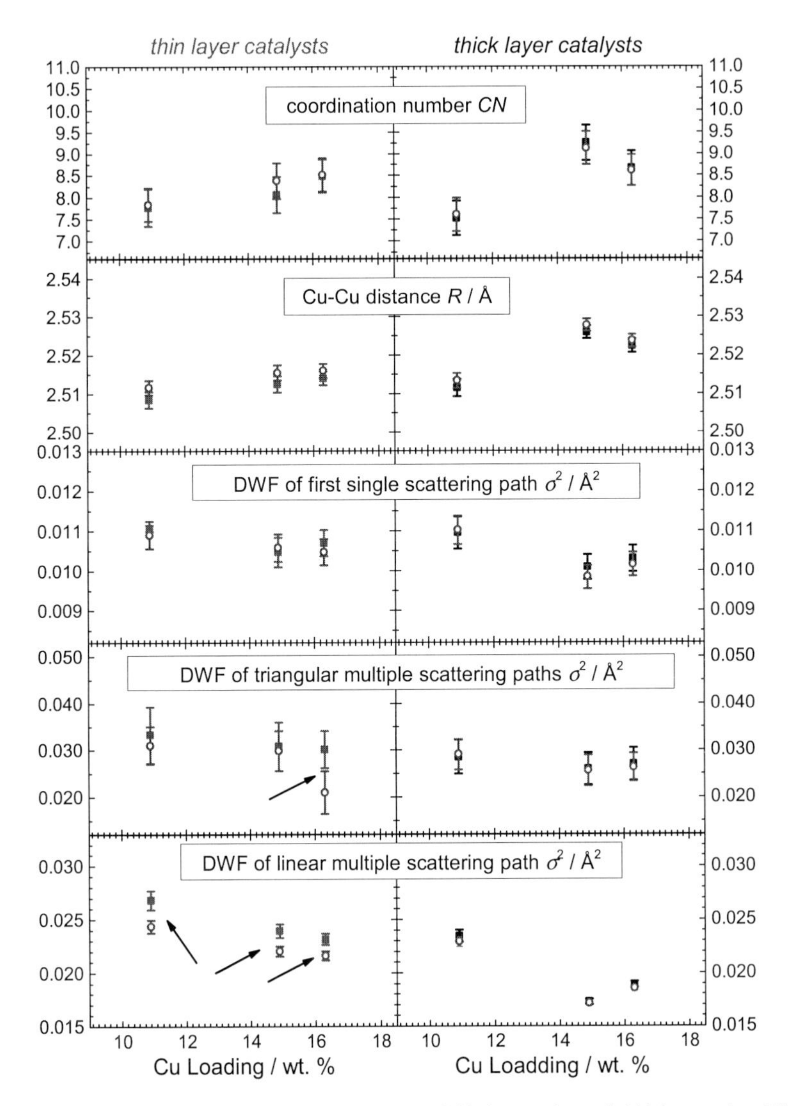

Figure III-16 Selection of refined structural parameters of *thin layer catalysts* and *thick layer catalysts*, left and right, respectively. Filled symbols represent Cu/SBA-15 catalysts after first reduction and empty symbols represent Cu/SBA-15 catalysts after second reduction. Debye-Waller-Factor (DWF) is the static disorder parameter. Errors are given with 95 % certainty.

The significant changes in the microstructure of CuO_x particles after the redox cycle suggested also changes in the local structure of Cu metal particles obtained by the second reduction after the redox cycle. However, the FT($\chi(k)\cdot k^3$) of EXAFS spectra of differently activated Cu/SBA-15 catalysts showed little differences (Figure III-15). With that, the EXAFS results agreed well with results obtained by XRD diffraction peak profile analysis. Re-reduced Cu metal particles after the redox cycle or after temporary oxygen co-feeding showed similar sizes but indicated less static disorder or microstrain.

III.3.11 Impact of redox activation on Cu surface areas

The Cu surface areas were determined after redox activation following the route described in *II.3.7*. In Figure III-17 the determined Cu surface areas of redox activated Cu/SBA-15 samples are compared to Cu surface areas of standard activated Cu/SBA-15 samples. Independent of the calcination mode, the redox activated Cu/SBA-15 samples exhibited larger Cu surface areas. Only the 16.3Cu *thin layer catalyst* and the 14.9Cu *thick layer catalyst* showed similar Cu surface areas after normal activation and after redox activation.

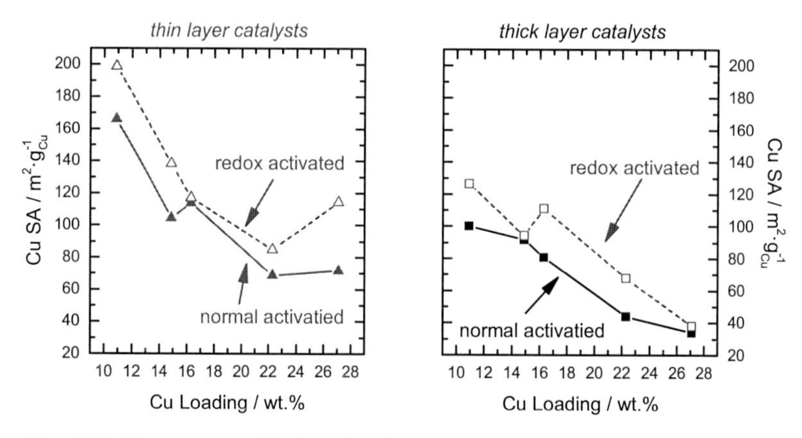

Figure III-17 Cu surface areas (SA) of standard activated Cu/SBA-25 catalysts (filled symbols) and redox activated Cu/SBA-15 catalysts (empty symbols). Cu surface areas of *thin layer* and *thick layer catalysts* are depicted on the left and right, respectively.

The increased Cu surface areas can be attributed to either reduced Cu particle sizes or modified Cu particle shapes. Increased Cu surface areas might be found, if more flat planes were present on Cu metal particles. Van der Grift *et al.*[120] reported a relationship between apparently larger Cu surface areas and shape of Cu metal particles. Both, the size of Cu metal particles and the constitution of Cu metal particles could therefore be affected during redox activation.

III.3.12 Methanol steam reforming after redox activation

Differently activated and differently calcined Cu/SBA-15 catalysts were tested in methanol steam reforming. The H_2 formation rates of Cu/SBA-15 catalysts are illustrated in Figure III-18 during 12 h time on stream. H_2 formation rates measured after standard activation are depicted as empty symbols. H_2 formation rates measured after redox activation are depicted as red symbols with a point in the center. Data for standard activated catalysts are depicted left and right for *thin layer* (green) and *thick layer catalysts* (black), respectively.

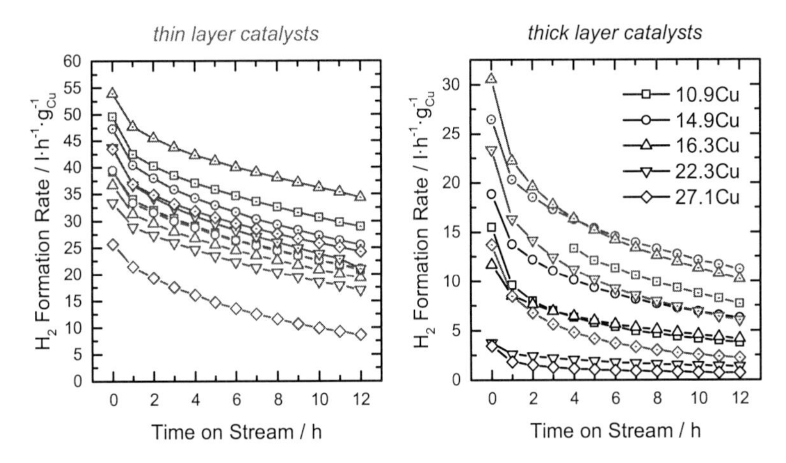

Figure III-18 H_2 formation rates after standard activation of *thin layer* and *thick layer catalysts* on the left (green, empty symbols) and on the right (black, empty symbols), respectively. Additionally, H_2 formation rates after redox activation of *thin layer* and *thick layer catalysts* are illustrated as corresponding red curves including points inside the symbols. Catalytic tests were applied at 250 °C in 2 % MeOH and 2 % H_2O balanced by He for 12 h time on stream.

The different catalytic performances of *thin layer* and *thick layer catalysts* were already discussed in *II.3.14*. Furthermore, the H_2 formation rates depended on the activation mode. Independent of calcination mode, all redox activated Cu/SBA-15 catalysts showed increased H_2 formation rates during 12 h time on stream in comparison to the corresponding normally activated Cu/SBA-15 catalysts. Interestingly, the *thick layer catalysts* exhibited enhanced activity increase due to redox activation, while the *thin layer catalysts* exhibited a moderate increase in activity after redox activation. This is illustrated in Figure III-19 as ratio between

H_2 formation rate after redox activation and H_2 formation rate after normal activation. In particular, the 22.3Cu and 27.1Cu *thick layer catalysts* formed distinctly more H_2 after redox activation.

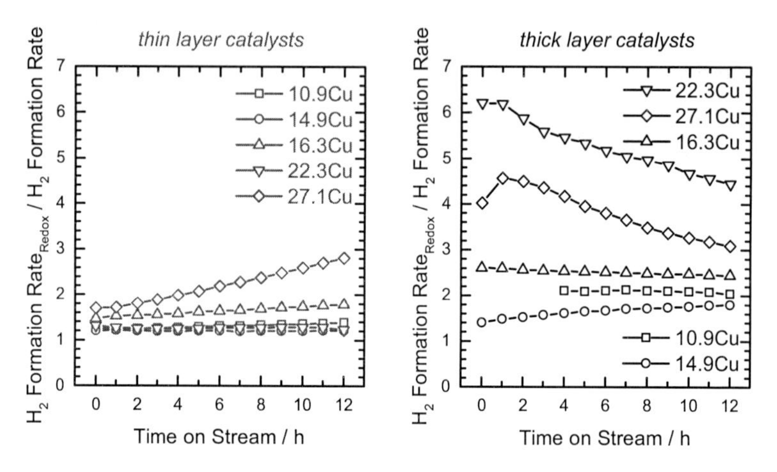

Figure III-19 Ratio between H_2 formation rate after redox activation and H_2 formation rate after normal activation of *thin layer catalysts* and *thick layer catalysts*, left and right, respectively.

Activity increase due to redox activation may be attributed to the previously found increased Cu surface areas. Therefore, the intrinsic activities needed to be compared. The intrinsic activities are summarized in Figure III-20 left for *thin layer catalysts* and on the right for *thick layer catalysts*.

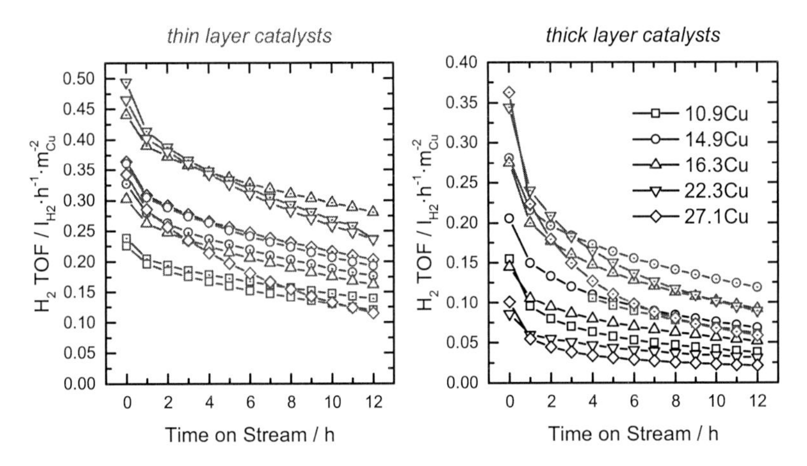

Figure III-20 H_2 turn over frequencies (TOFs) over 12 h time on stream of *thin layer catalysts* (left)and *thick layer catalysts* (right). Empty green and empty black symbols represent TOFs of normal activated catalysts and the corresponding redox activated catalysts are depicted as symbols with point in the center in red. H_2 was produced over Cu/SBA-15 catalysts at 250 °C in a stream of 2 % MeOH and 2 % H_2O balanced by He.

Thin layer and *thick layer catalysts* were differently affected by the redox activation. *Thin layer catalysts* retained their H_2 TOFs, except for 16.3Cu showing a moderately increased H_2 TOF. Conversely, *thick layer catalysts* produced more H_2 per Cu surface atom, if the *thick layer catalysts* were redox activated. Increased H_2 TOFs were thus observed, depending not only on calcination mode (compare *Chapter II*) but also on activation mode.

The ratio of H_2 TOFs after redox activation and H_2 TOFs after standard activation is depicted in Figure III-21. The *thin layer catalysts* showed a ratio of about 1. The 16.3Cu *thin layer catalyst* deviated with a ratio of 1.5. Within 12 h time on stream, the 27.1Cu *thin layer catalyst* exhibited little increase. This indicated that deactivation of 27.1Cu *thin layer catalysts* depended on activation mode. The redox activation apparently led to more stable Cu metal particles supported on SBA-15 that deactivated more slowly.

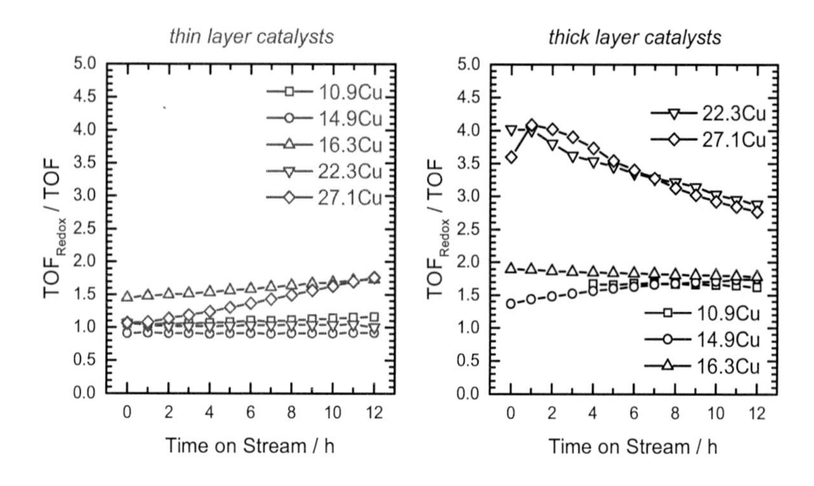

Figure III-21 Ratio between H_2 TOFs after redox activation and H_2 TOFs after normal activation of *thin layer catalysts* (left) and *thick layer catalysts* (right) during 12 h time on steam in methanol steam reforming feed at 250 °C.

The ratios of H_2 TOFs measured after redox activation and the H_2 TOFs measured after standard activation of *thick layer catalysts* were higher than 1. This ratio of 10.9Cu, 14.9Cu and 16.3Cu *thick layer catalysts* was located in the range between 1.5 and 2. 22.3Cu and 27.1Cu *thick layer catalysts* revealed ratios of about 4 at the beginning of methanol steam reforming, decreasing regularly to about 2.8 after 12 h time on stream. The deactivation of 22.3Cu and 27.1Cu *thick layer catalysts* proceeded apparently faster after redox activation. *Thick layer catalysts* revealed increased intrinsic activity after redox activation compared to standard activation, while the intrinsic activity of *thin layer catalysts* could not be improved by redox activation.

III.3.13 Evolution of Cu metal particles during redox activation

Cu metal exhibits a pronounced affinity to oxygen and easily forms an oxidic layer at the surface of Cu metal particles.[159] Further oxidation is governed by migrating Cu ions. Consequently, the following process illustrated in Figure III-22 may proceed during redox cycle. During first reduction, CuO particles supported in SBA-15 (a) were reduced to Cu metal particles (b). During re-oxidation, initially Cu_2O was formed at the surface (c). For further oxidation, Cu ions diffuse from the inner core to the particle surface leading to hollow spheres.[158,161] Furthermore, CuO was formed at the surface, and interfaces between the Cu, Cu_2O, and CuO were created (d). Strain at the interfaces evoked by the Kirkendall effect[162] resulted in collapsing hollow spheres. The formed smaller CuO particles (e) were transformed to smaller Cu metal particles (f) during the following reduction. These smaller Cu metal particles were prevented from sintering by interacting with the support.

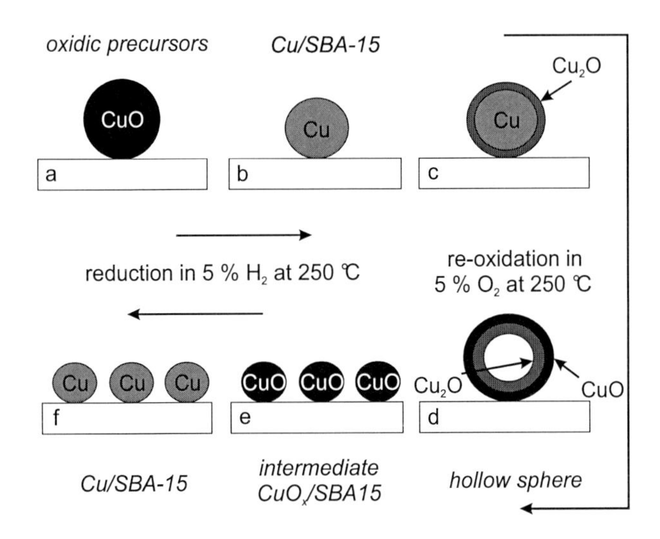

Figure III-22 Transformation of CuO_x particles during redox activation. CuO_x particles of oxidic precursors (a) were reduced to Cu metal particles (b). Re-oxidation starts at the particle surface (c) and leading to CuO_x hollow spheres (d). After collapsing, smaller CuO_x particles were present in the intermediate (e). Second reduction led to small Cu metal particles.

The same behavior was observed for supported Co particles during redox treatment.[158] Subsequent reduction led therefore to smaller Co metal particles. Consequently, the measured Cu surface areas of Cu/SBA-15 catalysts investigated here increased after the second reduction compared to once reduced Cu/SBA-15 catalysts. However, a corresponding decrease of Cu particle size was not detected. Compared to once reduced Cu metal particles, only less decrease of static disorder in twice reduced Cu metal particles was observed. Two effects may explain this apparent inconsistency. First, the increased Cu surface area based on re-dispersion of smaller Cu metal particles which cannot be detected by applied analytical technics. Second, the shape of Cu particle and the surface of Cu metal particles were mainly affected by redox activation.

The first assumption comprises Cu metal particles being smaller than 2 nm, because applied XRD analysis was limited. More detailed information may be achieved applying pair distribution function analysis of XRD patterns.[163] Dispersion of smaller Cu metal particles may therefore cause the observed increase of Cu surface areas. XAS analysis should reveal the reduction of particle sizes due to re-dispersion. Since this had not been observed, only a minor part of Cu metal particles was apparently transformed into smaller Cu metal particles leading to the observed increase of Cu surface areas and thus increased H_2 formation rates.

The second assumption comprises the varying contributions of Cu metal particle faces to Cu surface areas and varying amounts of surface defect structures. N_2O prefers flat planes and low index particle faces with regular surface structures to be chemisorbed dissociatively.[120] Changes of either one or both structural properties may affect the N_2O uptake on Cu metal particles and thus the apparent Cu surface area, although the actual particle size was unaffected. During redox activation, Cu metal particles were completely oxidized to CuO_x particles. During oxygen co-feeding Cu metal particles were partly oxidized to Cu/CuO_x particles. Then, both types of oxidized particles were re-reduced to Cu metal particles. During these transformations, the atoms of Cu metal particles were rearranged. Hence, the weighting of newly formed particles faces[120], the amount and types of defects, and stacking faults on the surface were able to change distinctly. As it has been previously shown[47], the support plays a crucial role at this point. For example, Cu metal particles on ZnO tended to form microstrain after temporary oxygen addition during MSR.[11] Cu metal particles supported on ZrO_2 did not re-reduce completely.[23] For the here investigated Cu/SBA-15 catalysts, there was still a lack of information on surface properties of Cu metal particles. Hence, the second assumption dealing with variation of Cu metal particle surfaces due to redox treatment cannot be supported unrestrictedly. However, the influence of surface planes of particles in heterogeneous catalysis[47,103,147,164] and the influence of defects in Cu metal particles in methanol chemistry[6] has been shown.

Both, the re-dispersion of small Cu metal particles and the variation of Cu metal particle surfaces could not be observed. However, both contribute to the catalytic activity. They were therefore likely to cause the observed increase of H_2 production occurring in *thin layer* and *thick layer catalysts*.

III.3.14 Activity in methanol steam reforming after redox treatment

Increasing H_2 formation rates due to re-dispersion during redox treatment can be observed while H_2 TOFs are invariant. For *thin layer catalysts*, the ratio of TOFs of redox activated and standard activated catalysts was almost one. Hence, the redox activation led to an increased number of active sites detected by dissociative N_2O chemisorption. The H_2 TOF at each reactions center was invariant.

In *thick layer catalysts*, additional increase of the intrinsic activity was observed. The ratio between H_2 TOFs of normal activated and redox activated catalysts was about two for 10.9Cu, 14.9Cu and 16.3Cu *thick layer catalysts* and four for 22.3Cu and 27.1Cu *thick layer catalysts*. It seemed that *thick layer catalysts* consisting of larger Cu metal particles exhibited pronounced increased intrinsic activity after redox activation compared to standard activated *thick layer catalysts*. This indicated that besides re-dispersion of Cu metal particles the TOF at each observed active site increased significantly. The increased intrinsic activities can be explained by disintegrating CuO nanoparticles due to redox activation described above in Figure III-22. The smaller Cu metal particles consisted of a higher fraction of edges, kinks, and corners, which were probably more active in methanol steam reforming (compare *II.3.14*).

The re-dispersion process apparently occurred only in larger Cu metal particles of 22.3Cu and 27.1Cu *thick layer catalysts*, which showed the largest increase in intrinsic activity. This process could also explain the increased activity after temporary oxygen co-feeding. Although the Cu metal particles were not completely oxidized to CuO particles during oxygen co-feeding, at least roughening of Cu metal particle surfaces was possible due to Cu atom movement. This may lead to higher fraction of edge atoms or defects on the Cu metal particle surfaces and thus to increased activities. Increased activity after oxygen co-feeding is often discussed as burning of adsorbed species or coke.[17,157] Both were unlikely due to use of purified feed, and Cu metal particles do not tend to coke formation in methanol steam reforming (compare *II.3.16*). Furthermore, the increased activity was also observed after redox activation while Cu catalysts did not contact organic compounds. This suggested that the observed increase of activity after redox pretreatment may be attributed to structural changes such as shape or microstrain of the Cu metal particles supported on SBA-15.

III.4 Conclusions

Cu/SBA-15 model catalysts exhibited an increased H_2 production in methanol steam reforming after oxygen co-feeding. The Cu metal particles of Cu/SBA-15 catalysts were partly oxidized. Extend of oxidation depended on the calcination mode of used catalysts and correlated with Cu metal particle sizes. Besides increased H_2 production less methyl formate was produced as sideproduct.

Furthermore, the structural changes and the changes of Cu surface areas were investigated by employing the redox activation procedure. The redox activation procedure was assumed to affect the Cu metal particles equally after oxygen co-feeding and after redox activation. After the redox cycle, the CuO_x particles of intermediate CuO_x/SBA-15 showed distinct deviation in local disorder compared to the initial oxidic precursors. The extent of structural changes was larger for *thin layer precursors* than for *thick layer precursors*. The increased local disorder in CuO_x particles supported on SBA-15 correlated with their increased reducibility observed during the second reduction of the redox activation procedure. Therefore, the redox activation affected mainly the constitution of CuO_x particles. Hence, the size of CuO_x particles on the support was apparently unaffected by the redox activation procedure, indicating strong interaction between CuO_x particles and the support.

The CuO_x particles of oxidic precursors and intermediate CuO_x/SBA-15 exhibited distinct structural differences. Reduction of these different CuO_x particles resulted in similar Cu metal particles supported on SBA-15. However, these Cu metal particles differed in Cu surface areas. Moreover, reduced static disorder in Cu metal particles of *thin layer catalysts* after redox activation was indicated by EXAFS analysis. Apparently, the structural characteristics of Cu metal particles were governed by the interaction between Cu metal particles and support.

Increased H_2 formation rates of Cu/SBA-15 catalysts were observed after employing a redox activation. The H_2 TOFs of *thick layer catalysts* were increased after redox activation while the H_2 TOFs of *thin layer catalysts* were similar. Hence, increased methanol steam reforming activity of *thin layer catalysts* was achieved by apparently increased dispersion due to redox activation. Increased catalytic activity of *thick layer catalysts* was achieved by both, re-dispersion and enhanced intrinsic activity of Cu metal particles. Re-dispersion led to smaller Cu metal particles. In *Chapter II* it was shown that smaller Cu metal particles exhibited increased intrinsic activities. Larger Cu metal particles may show higher dispersions according to the process described in Figure III-22. Therefore, the strongest increase in H_2 TOFs was observed for 22.3Cu and 27.1Cu *thick layer catalysts* showing largest Cu metal particles after standard activation. Despite the latter two catalysts, the increase in H_2 formation rates and H_2 TOFs was permanently over a period of 12 h. After redox cycle, a stronger interaction of Cu metal particles and the support might be present leading to more stable Cu metal particles.

The redox cycle might also evoke roughening of Cu metal surfaces leading to more unsaturated Cu surface atoms. These unsaturated Cu atoms would show increased intrinsic activity. Altogether, a certain Cu metal particle size was required to observe an increase of intrinsic MSR activity of Cu/SBA-15 catalyst after redox treatment. The extent of increase rose with increasing Cu metal particle size. This in combination with only little change in disorder of Cu metal particles indicated size related H_2 TOFs of Cu metal particles in MSR. More detailed information on shape of Cu metal particles may be obtained from SAXS measurements. Further investigation on Cu metal particle sizes and Cu metal particle distributions using analytical methods such as XAS, XRD, or TEM, would maybe complement the observed relationships.

General conclusions and outlook

CuO_x/SBA-15 model system as oxidic precursor

Synthesis of oxidic precursors succeeded under preservation of the mesoporous structure of the support material. Using the citrate route, varying the Cu loading, and varying the layer thickness of sample powders during calcination yielded various CuO_x particles on SBA-15. In *thick layer precursors*, CuO_x particles were found to be arranged in mesopores. In *thin layer precursors*, amorphous CuO_x particles clocked the micropores in pore walls of SBA-15. Structure analysis indicated small and uniform CuO_x particles in *thin layer precursors*, while structurally ordered CuO_x particles in *thick layer precursors* depended on Cu loading and were larger compared to those in corresponding *thin layer precursors*. It can be assumed that diffusion barriers during calcination hindered removing gaseous products of decomposition. Hence, the gaseous products may longer interact with Cu ions in *thick layer precursors* increasing the Cu ion mobility and thus, enabling CuO particle growth. Hence, CuO_x particles were differently dispersed on the support depending on calcination mode. CuO_x particles of *thin layer precursors* showed enhanced reducibility compared to those in *thick layer precursors*. The simple access to various structures and various sizes of CuO_x particles make the applied preparation route interesting for further investigation on supported metal oxide model catalysts. According to the concept of *chemical memory effect*, the different precursors were a promising basis to create various metal particles supported on SBA-15 by reducing metal oxide particles to corresponding metals.

Cu/SBA-15 model catalysts in methanol steam reforming

Here, Cu metal particles varying in structure and size were accessible by varying the calcination mode. The structural differences of CuO_x particles present in oxidic precursors were preserved during activation. Hence, *thin layer catalysts* showed more dispersed Cu metal particles than *thick layer catalysts*. Cu metal particles present in *thin layer catalysts* showed similar sizes and larger Cu surface areas, while those in *thick layer catalysts* varied in size and exhibited smaller Cu surface areas. Moreover, the small Cu metal particles in *thin layer catalysts* showed considerable microstrain and increased disorder compared to those in *thick layer catalysts*. It was concluded that microstrain was size related and that the interface between Cu metal particles and support may be responsible for either the stabilization of small particles or the microstrain in the small particles. The support apparently stabilized smaller Cu metal particles more than larger Cu metal particles.

In accordance to structural difference, the *thin layer catalysts* showed larger activities in methanol steam reforming compared to the corresponding *thick layer catalysts*. Both, increased H_2 formation rates and increased H_2 TOFs were observed employing *thin layer catalysts*. The increased activity correlated with increased disorder in Cu metal particles present in *thin layer catalysts*. Moreover, the catalytic activity was governed by Cu metal particle size and was independent of performed calcination mode during Cu/SBA-15 synthesis. Cu nanoparticles being smaller than 3 nm exhibited increased H_2 formation rates and increased H_2 TOFs. Enhanced H_2 formation was attributed to the increasing number of kinks, edges, and corners in particles of reduced size. Therefore, a correlation between constitution of Cu metal particles and activity in methanol steam reforming could be drawn.

Interestingly, the smaller Cu metal particles of *thin layer catalysts* deactivated less than those of *thick layer catalysts*, underlining the stabilizing effect of the support on Cu metal particles. During methanol steam reforming at 250 °C no sintering of Cu metal particles was observed.

The activity of Cu /SBA-15 catalysts could be increased by either temporarily adding oxygen to the methanol steam reforming feed or by employing redox cycle before activation (redox activation). The enhanced H_2 production after redox activation was constant over a period of 12 h agreeing well with findings that silica stabilized Cu metal particles. During redox pretreatment CuO_x particles of oxidic precursors were transformed into CuO nanoparticles present in intermediate CuO_x/SBA-15 samples. These CuO nanoparticles exhibited increased local disorder. Hence, reduction of the CuO nanoparticles was observed at lower temperatures than reduction of initial CuO_x particles. A good correlation between static disorder and reducibility was found. Therefore, the redox pretreatment did not affect the size of CuO_x particles present in oxidic precursors or in intermediate CuO_x/SBA-15 samples but influenced the local structure.

Similarly to preserved sizes of CuO_x particles, the size of Cu metal particles was unaffected by redox pretreatment or oxygen co-feeding. However, re-reduced Cu metal particles exhibited enlarged Cu surface areas and indicated less disorder after redox treatment. Cu metal particles showing structural similarity were obtained from CuO_x particles that had exhibited considerable structural differences. Apparently, not only the local structure of CuO_x particles decided on structure of Cu metal particles after activation but also the interaction between particles and support may be decisive for the structure of Cu metal particles.

The observed increase of Cu surface areas due to redox activation may be attributed to changes in the surface structure of Cu metal particles potentially evoked by interaction between Cu metal particles and silica support. Remarkably, the catalytic activity of the largest Cu metal particles of Cu/SBA-15 catalysts showed after redox pretreatment the highest increase in methanol steam reforming activity. Apparently, a certain Cu metal particle size was required to observe Cu metal particles of reduced size after redox treatment. The smaller Cu metal particles were probably well stabilized because Cu metal particles interacted strongly with the support.

Microstrain in Cu metal particles

Considerable microstrain in Cu metal particles present in *thin layer catalysts* was detected without any interface to ZnO. This finding leads to further questions about the role of ZnO as promotor in Cu/ZnO catalysts with respect to their catalytic activities. It would be interesting to deposit Cu nanoparticles of the same size as here reported in combination with ZnO on SBA-15 considering the optimal Cu/Zn ratio of 7/3. The structural characteristics of Cu/ZnO/SBA-15 catalysts may reveal the influence of ZnO on the structure of Cu metal particles exhibiting the same size and the same chemical composition as well as the influence of ZnO on the catalytic activity in methanol steam reforming.

Moreover, the reasons for microstrain in Cu metal particles supported on SBA-15 could not be elucidated. The investigations indicated that the Cu/SBA-15 interface may play a crucial role. Therefore, further investigations would be interesting applying other analytical methods such as Ar physisorption, SAXS, and ^{63}Cu-NMR. Ar physisorption measurements may contribute to understand the interaction between silica and Cu metal particles. Especially the position of Cu metal particles may be elucidated (extent of micropore clocking) leading to better estimations of Cu/silica interface. SAXS measurement may give information about the shapes of Cu metal particles and the corresponding changes evoked by redox treatment. The ^{63}Cu-NMR spectra may distinguish the extent of microstrain in Cu metal particles observed in *thin layer* and *thick layer catalysts* in addition to the here presented *in situ* XRD and *in situ* XAS results.

Appendix

DR-UV/Vis

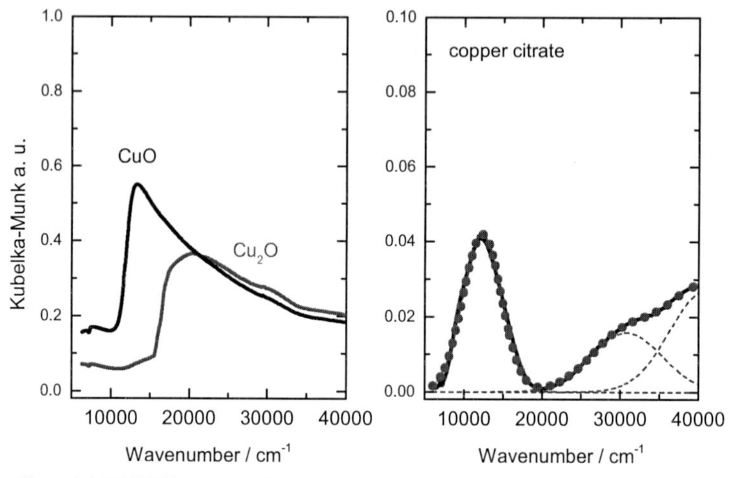

Figure 0-1 DR-UV/Vis spectra of crystalline CuO and Cu_2O (left) and crystalline Cu citrate (right). Theoretical spectrum of Cu citrate is added (right, blue dots), which results from three Gaussian functions given as blue broken lines.

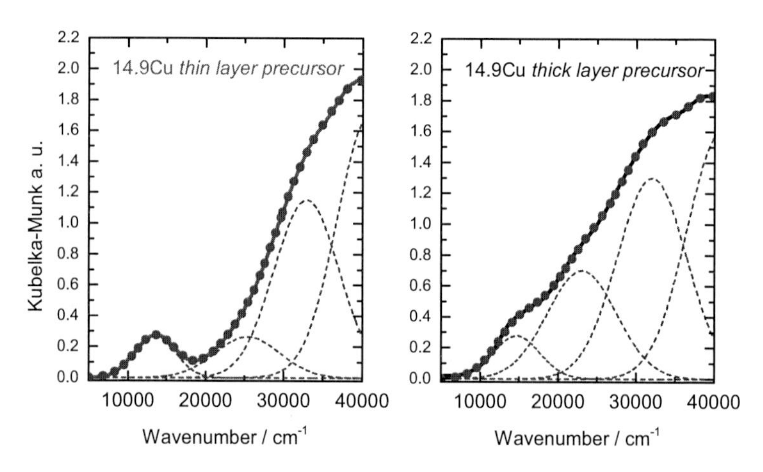

Figure 0-2 DR-UV/Vis spectra of 14.9Cu *thin layer precursor* (left, green) and 14.9Cu *thick layer precursor* (right, black). Blue broken lines represent fitted Gaussian functions of single contributions of each absorption band and resulting spectra (blue dots).

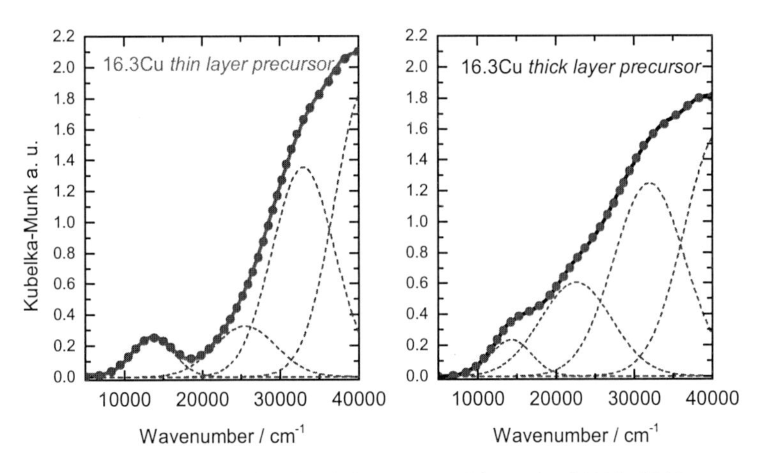

Figure 0-3 DR-UV/Vis spectra of 16.3Cu *thin layer precursor* (left, green) and 16.3Cu *thick layer precursor* (right, black) 16.3Cu. Blue broken lines represent fitted Gaussian functions of single contributions of each absorption band and resulting spectra (blue dots).

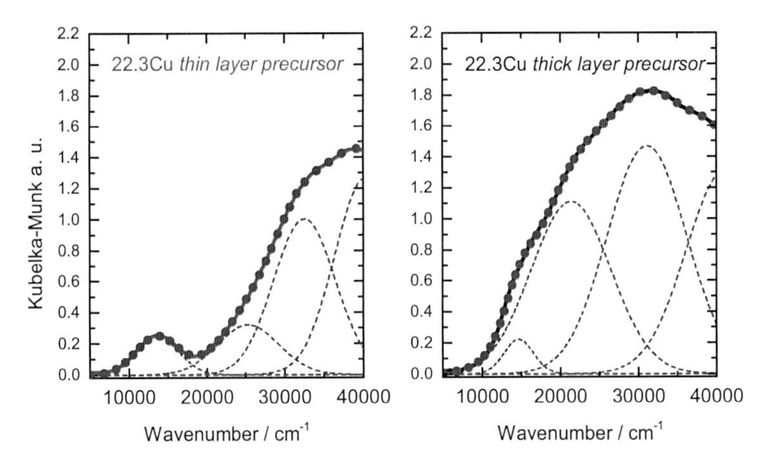

Figure 0-4 DR-UV/Vis spectra of 22.3Cu *thin layer precursor* (left, green) and 22.3Cu *thick layer precursor* (right, black). Blue broken lines represent fitted Gaussian peaks of single contributions of each absorption band and resulting spectra (blue dots).

XAS refinements of oxidic precursors

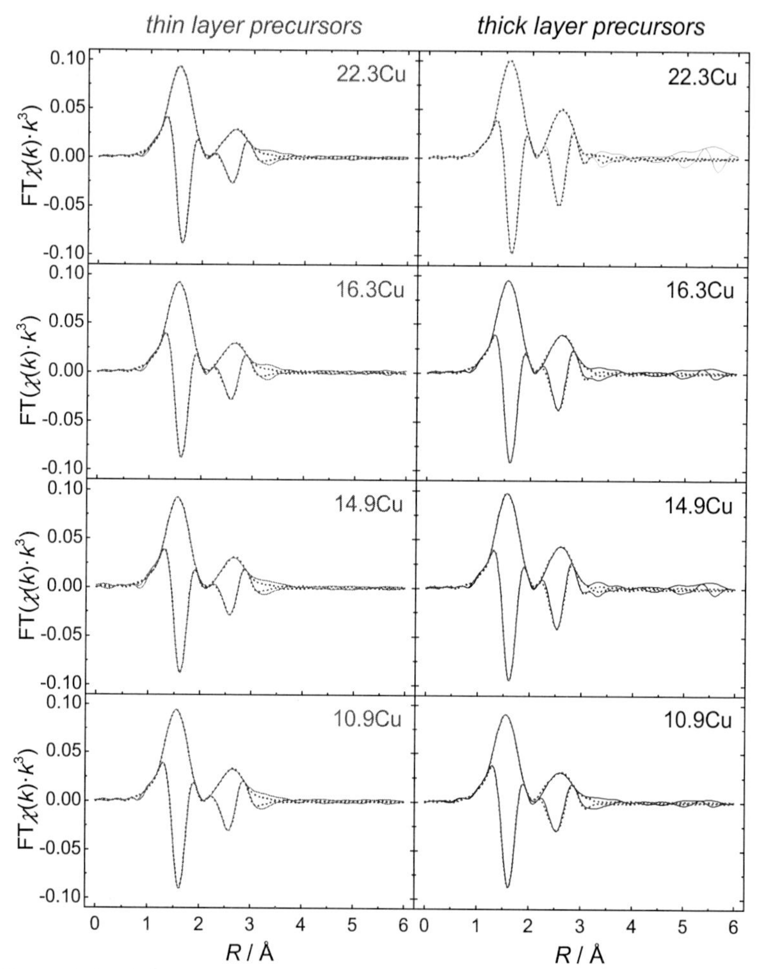

Figure 0-5 FT($\chi(k)\cdot k^3$) and the corresponding imaginary part of *thin layer precursors* (left) and *thick layer precursors* /right) in green and in black, respectively. Calculated FT($\chi(k)\cdot k^3$) and the corresponding imaginary part are shown as blue dots. Cu loading of oxidic precursors is given as *number* before *Cu* top right of each diagram.

Table 0–1 EXAFS refinement results of *thin layer precursors* on basis of CuO structure. Parameters were optimized independently. Errors were calculated assuming 95 % certainty. [a] Consisted of two Cu-O backscattering paths. Each path showed *CN* of 2, and variation of distances was kept constant. [a] Only the results of first scattering path are listed. [b] *CN* was set to 1. [c] *CN* was set to 2.

	10.9Cu	14.9Cu[c]	16.3Cu	22.3Cu[c]	27.1Cu[c]	CuO
R (Cu-O)[a] / Å	1.942 ± 0.001	1.945 ± 0.006	1.9455 ± 0.006	1.945 ± 0.005	1.946 ± 0.005	1.96 ± 0.015
R (Cu-O)[b] / Å	2.77 ± 0.015	2.77 ± 0.018	2.77 ± 0.022	2.76 ± 0.014	2.77 ± 0.016	2.77 ± 0.023
σ^2 (Cu-O) / 10^{-3} Å^2	4.8 ±0.1	4.9 ± 0.1	4.9 ± 0.1	4.8 ± 0.1	4.8 ± 0.1	3.9 ± 0.2
R (Cu-Cu)[b] / Å	2.952 ± 0.004	2.955 ± 0.004	2.956 ± 0.004	2.955 ± 0.004	2.953 ± 0.004	2.914 ± 0.004
σ^2 / 10^{-3} Å^2	9.9 ± 0.29	10.0 ± 0.35	10.0 ± 0.44	10.2 ± 0.30	10.2 ± 0.33	4.5 ± 0.24
E_0 / eV	4.0502	4.0742	4.1648	3.5953	3.551	8.105
R	4.69	4.96	6.07	4.79	4.95	7.33

Table 0–2 EXAFS refinement results of *thick layer precursors* on basis of CuO structure. Parameters were optimized independently, and errors were calculated assuming 95 % certainty. [a] Consisted of two Cu-O backscattering paths. Each path showed a *CN* of 2 and variation of distance was kept constant. [a] Only the results of 1st scattering path are listed. [b] *CN* was set to 2. [c] *CN* was set to 0.5. [d] *CN* was set to 3 [e] *CN* was set to 1.5.

	10.9Cu	14.9Cu	16.3Cu	22.3Cu	27.1Cu	CuO
R (Cu-O)[a] / Å	1.944 ± 0.007	1.944 ± 0.007	1.942 ± 0.001	1.947 ± 0.001	1.944 ± 0.002	1.96 ± 0.015
R (Cu-O) / Å	2.79 ± 0.01	2.79 ± 0.01	2.79 ± 0.01	2.79 ± 0.01	2.81 ± 0.03	2.77 ± 0.02
σ^2 (Cu-O) / 10^{-3} Å^2	5.09 ± 0.1	4.51 ± 0.09	4.73 ± 0.08	4.32 ± 0.09	4.89 ± 0.12	3.9 ± 0.2
R (Cu-Cu) / Å	2.958 ± 0.006[b]	2.941 ± 0.004[b]	2.942 ± 0.004[b]	2.936 ± 0.004[d]	2.937 ± 0.003[d]	2.913 ± 0.005
R (Cu-Cu) / Å	2.958 ± 0.006[c]	3.14 ± 0.02[c]	3.14 ± 0.02[c]	3.147 ± 0.008[e]	3.14 ± 0.01[e]	3.064 ± 0.005
σ^2 / 10^{-3} Å^2	8.7 ± 0.3[b]	6.6 ± 0.2[b]	7.2 ± 0.2[b]	7.5 ± 0.2[c]	8.3 ± 0.2[c]	4.5 ± 24
E_0 / eV	4.75	5.31	4.79	6.15	6.08	8.105
R	4.25	3.42	3.18	3.43	4.16	7.33

Temperature-programmed reduction (TPR)

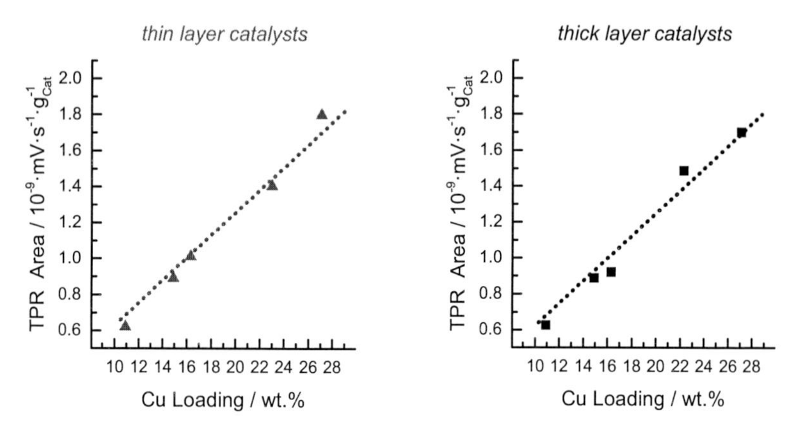

Figure 0-6 Correlation between H_2 consumption peak integral and Cu loading of Cu/SBA-15 catalysts. Dotted line represents linear fit forced through the origin.

Table 0–3 Calculate *K* and *P* achieved in TPR experiments using 30 mg sample and a flow of 40 ml/min of 5 % H_2 balanced by Ar.

	10.9Cu	14.9Cu	16.3Cu	22.3Cu	27.1Cu
K / s	34.6	47.3	51.7	70.8	86.0
P / K	1.9	3.9	4.3	5.9	7.2

Sideproducts in methanol steam reforming

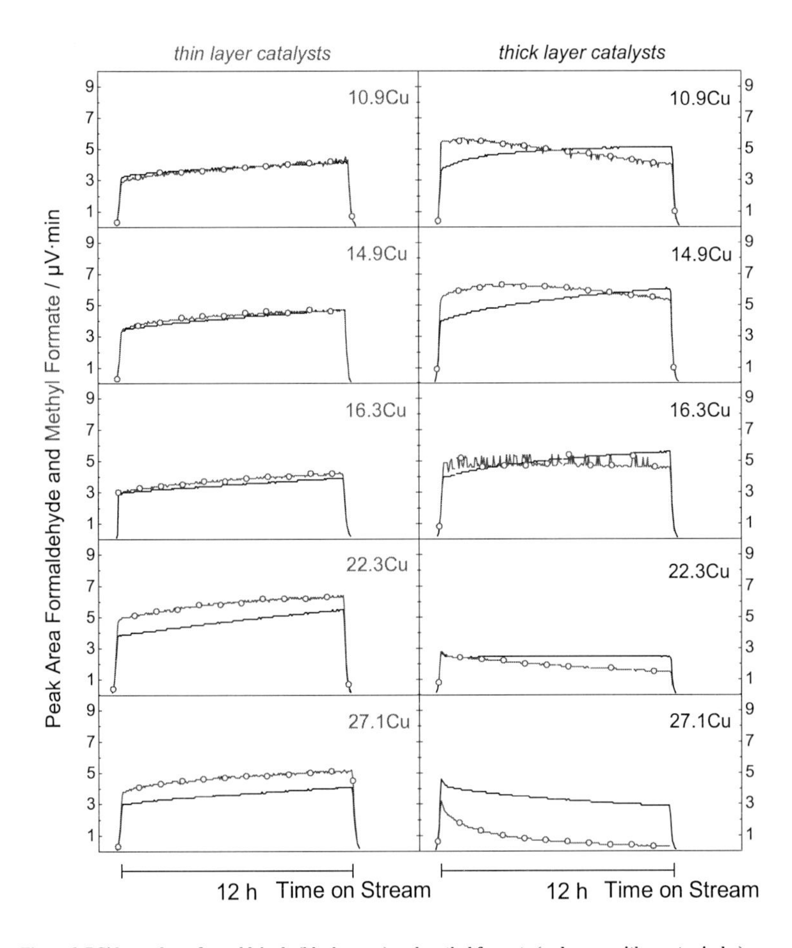

Figure 0-7 Side products formaldehyde (black curve) and methyl formate (red curve with empty circles) over *thin layer* and *thick layer catalysts* left and right, respectively. Peak area is given as integral of corresponding peak in gas chromatogram.

XAS during oxygen co-feeding

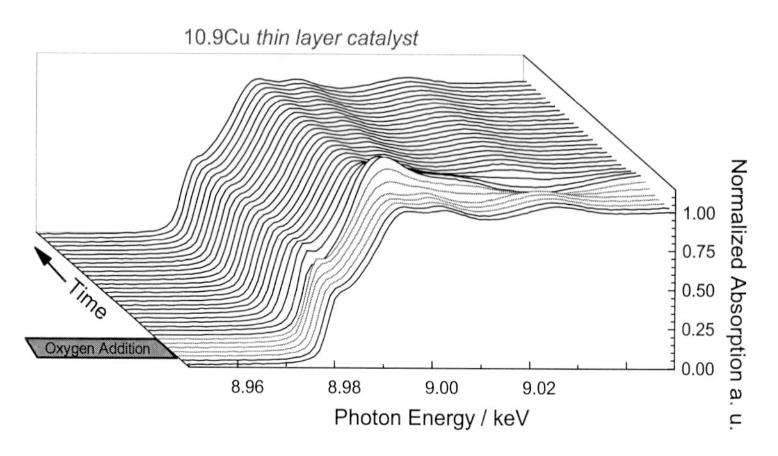

Figure 0-8 Evolution of XANES spectra of 10.9Cu *thin layer catalyst* at Cu K edge during methanol steam reforming at 250 °C in a stream of 30 ml/min of 2 % MeOH and 2 % H_2O balanced by He and employed temporary O_2 addition of 5 % in a total stream of 40 ml/min. Temporary O_2 addition is marked as orange area and orange spectra.

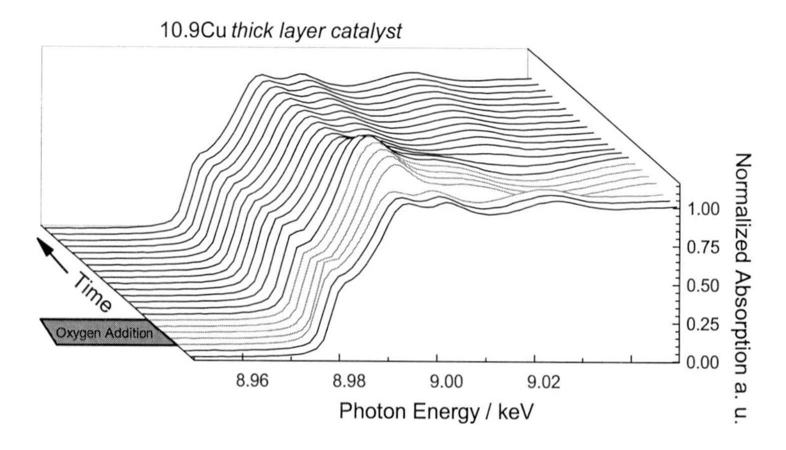

Figure 0-9 Evolution of XANES spectra of 10.9Cu *thick layer catalyst* at Cu K edge during methanol steam reforming at 250 °C in a stream of 30 ml/min of 2 % MeOH and 2 % H_2O balanced by He and employed temporary O_2 addition of 5 % in a total stream of 40 ml/min. Temporary O_2 addition is marked as orange area and orange spectra.

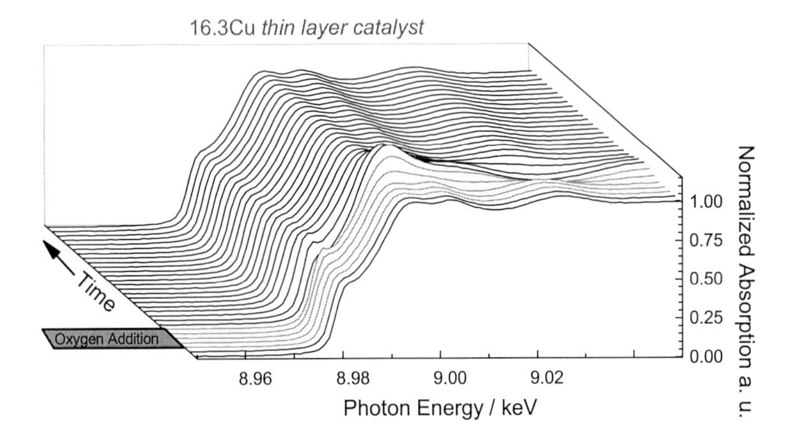

Figure 0-10 Evolution of XANES spectra of 16.3Cu *thin layer catalyst* at Cu K edge during methanol steam reforming at 250 °C in a stream of 30 ml/min of 2 % MeOH and 2 % H_2O balanced by He and employed temporary O_2 addition of 5 % in a total stream of 40 ml/min. Temporary O_2 addition is marked as orange area and orange spectra.

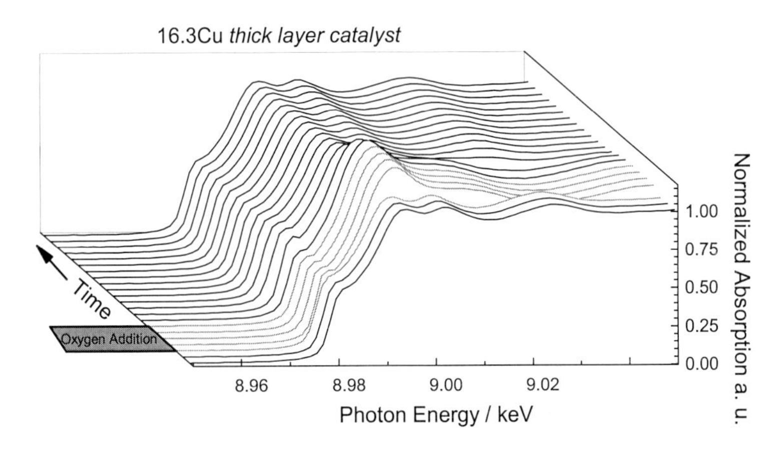

Figure 0-11 Evolution of XANES spectra of 16.3Cu *thick layer catalyst* at Cu K edge during methanol steam reforming at 250 °C in a stream of 30 ml/min of 2 % MeOH and 2 % H_2O balanced by He and employed temporary O_2 addition of 5 % in a total stream of 40 ml/min. Temporary O_2 addition is marked as orange area and orange spectra.

Bibliography

[1] S. K. Kamarudin, N. Shariah shamsul, *Aust. J. Basic Appl. Sci.* **2014**, *8*, 81–83.

[2] S. Zander, E. L. Kunkes, M. E. Schuster, J. Schumann, G. Weinberg, D. Teschner, N. Jacobsen, R. Schlögl, M. Behrens, *Angew. Chem. Int. Ed. Engl.* **2013**, *52*, 6536–6540.

[3] G. A. Olah, A. Goeppert, Prakash, G. K. Surya, *Beyond oil and gas. The methanol economy,* 2nd updated and enlarged ed. ed., Wiley-VCH, Weinheim, **2009**.

[4] P. de Wild, M. J. F. M. Verhaak, *Catalysis Today* **2000**, *60*, 3–10.

[5] S. Sã, H. Silva, L. Brandão, J. M. Sousa, A. Mendes, *Appl. Catal., B* **2010**, *99*, 43–57.

[6] M. Behrens et al., *Science* **2012**, *336*, 893–897.

[7] G. C. Chinchen, P. J. Denny, J. R. Jennings, M. S. Spencer, K. C. Waugh, *Appl. Catal.* **1988**, *36*, 1–65.

[8] M. Behrens, S. Zander, P. Kurr, N. Jacobsen, J. Senker, G. Koch, T. Ressler, R. W. Fischer, R. Schlögl, *J. Am. Chem. Soc.* **2013**, *135*, 6061–6068.

[9] W. J. Moore, D. O. Hummel, G. Trafara, *Physikalische Chemie,* 4., durchges. und verb. Aufl. ed., de Gruyter, Berlin [u.a.], **1986**.

[10] P. Kurr, I. Kasatkin, F. Girgsdies, A. Trunschke, R. Schlögl, T. Ressler, *Appl. Catal., A* **2008**, *348*, 153–164.

[11] M. M. Günter, T. Ressler, R. E. Jentoft, B. Bems, *J. Catal.* **2001**, *203*, 133–149.

[12] B. L. Kniep, T. Ressler, A. Rabis, F. Girgsdies, M. Baenitz, F. Steglich, R. Schlögl, *Angew. Chem. Int. Ed. Engl.* **2004**, *43*, 112–115.

[13] K. Takahashi, H. Kobayashi, N. Takezawa, *Chem. Lett.* **1985**, *14*, 759–762.

[14] D. R. Palo, R. A. Dagle, J. D. Holladay, *Chem. Rev.* **2007**, *107*, 3992–4021.

[15] F. Liao, Y. Huang, J. Ge, W. Zheng, K. Tedsree, P. Collier, X. Hong, S. C. Tsang, *Angew. Chem. Int. Ed. Engl.* **2011**, *50*, 2162–2165.

[16] Y. Matsumura, H. Ishibe, *J. Catal.* **2009**, *268*, 282–289.

[17] G. Ertl, H. Knözinger, F. Schüth, J. Weitkamp (Eds.) *Handbook of heterogeneous catalysis,* 2., completely rev. and enl. ed. ed., Wiley-VCH-Verlagsgemeinschaft, Weinheim.

[18] M. Behrens, R. Schlögl, *Z. Anorg. Allg. Chem.* **2013**, *639*, 2683–2695.

[19] M. M. Günter, T. Ressler, B. Bems, C. Büscher, T. Genger, O. Hinrichsen, M. Muhler, R. Schlögl, *Catal. Lett.* **2001**, *71*, 37–44.

[20] S. Zander, B. Seidlhofer, M. Behrens, *Dalton transactions* **2012**, *41*, 13413–13422.

[21] C. Yao, L. Wang, Y. Liu, G. Wu, Y. Cao, W. Dai, H. He, K. Fan, *Appl. Catal., A* **2006**, *297*, 151–158.

[22] B. S. Clausen, J. Schiøtz, L. Gråbæk, C. V. Ovesen, K. W. Jacobsen, J. K. Nørskov, H. Topsøe, *Top. Catal.* **1994**, *1*, 367–376.

[23] A. Szizybalski, F. Girgsdies, A. Rabis, Y. Wang, M. Niederberger, T. Ressler, *J. Catal.* **2005**, *233*, 297–307.

[24] R. E. Dinnebier, S. Billinge, *Powder Diffraction: Theory and Practice*, Royal Society of Chemistry, **2008**.

[25] G. Koch, K. Meurisch, T. Ressler, *Z. Anorg. Allg. Chem.* **2010**, *636*, 2094.

[26] C.-H. Liu, N.-C. Lai, J.-F. Lee, C.-S. Chen, C.-M. Yang, *J. Catal.* **2014**, *316*, 231–239.

[27] C. Marcilly, P. Courtey, B. Delmon, *J. Am. Cer. Soc.* **1970**, *53*, 56–57.

[28] M.-H. Looi, S.-T. Lee, S. B. Abd-Hamid, *Chin. J. Catal.* **2008**, *29*, 566–570.

[29] a) Gottwald, *GC für Anwender*, Wiley-VCH-Verlagsgemeinschaft; Weinheim, **1995**; b) B. Baars, H. Schaller, *Fehlersuche in der Gaschromatographie. Diagnose aus dem Chromatogramm*, VCH, Weinheim, New York, **1994**;

[30] L. Spieß, *Moderne Röntgenbeugung. Röntgendiffraktometrie für Materialwissenschaftler, Physiker und Chemiker,* 2., überarb. u. erw. Aufl. ed., Teubner, Wiesbaden, **2008**.

[31] D. A. Skoog, J. J. Leary, *Instrumentelle Analytik. Grundlagen, Geräte, Anwendungen*, Springer, Berlin [u.a.], **1996**.

[32] a) S. D. Kelly, D. Hesterberg, B. Ravel in *Soil Science Society of America book series, no. 5* (Eds.: A. L. Ulery, L. R. Drees), Soil Science Society of America, Madison, WI, **2008**; b) D. C. Koningsberger, R. Prins, *X-ray absorption. Principles, applications, techniques of EXAFS, SEXAFS, and XANES*, Wiley, New York, **1988**, *v. 92*;

[33] A. Mansour, P. Smith, W. Baker, M. Balasubramanian, J. McBreen, *Electrochimica Acta* **2002**, *47*, 3151–3161.

[34] T. Ressler, J. Wienold, R. E. Jentoft, T. Neisius, *J. Catal.* **2002**, *210*, 67–83.

[35] M. Fernández-García, *Catal. Rev.* **2002**, *44*, 59–121.

[36] D. Bazin, J. J. Rehr, *J. Phys. Chem. B* **2003**, *107*, 12398–12402.

[37] D. C. Koningsberger, B. L. Mojet, G. E. van Dorseen, D. E. Ramaker, *Top. Catal.* **2000**, *10*, 143–155.

[38] G. Kortüm, *Reflexionsspektroskopie. Grundlagen, Methodik, Anwendungen*, Springer Berlin Heidelberg; Imprint: Springer, Berlin, Heidelberg, **1969**.

[39] Z. Sojka, F. Bozon-Verduraz, M. Che in *Handbook of heterogeneous catalysis*. 2., completely rev. and enl. ed. ed. (Eds.: G. Ertl, H. Knözinger, F. Schüth, J. Weitkamp), Wiley-VCH-Verlagsgemeinschaft, Weinheim.

[40] A. V. Neimark, K. S. W. Sing, M. Thommes in *Handbook of heterogeneous catalysis*. 2., completely rev. and enl. ed. ed. (Eds.: G. Ertl, H. Knözinger, F. Schüth, J. Weitkamp), Wiley-VCH-Verlagsgemeinschaft, Weinheim.

[41] S. Brunauer, P. H. Emmett, E. Teller, *J. Am. Chem. Soc.* **1938**, *60*, 309–319.

[42] E. P. Barrett, L. G. Joyner, P. P. Halenda, *J. Am. Chem. Soc.* **1951**, *73*, 373–380.

[43] H. Knözinger in *Handbook of heterogeneous catalysis*. 2., completely rev. and enl. ed. ed. (Eds.: G. Ertl, H. Knözinger, F. Schüth, J. Weitkamp), Wiley-VCH-Verlagsgemeinschaft, Weinheim.

[44] G. Bergeret, P. Gallezot in *Handbook of heterogeneous catalysis*. 2., completely rev. and enl. ed. ed. (Eds.: G. Ertl, H. Knözinger, F. Schüth, J. Weitkamp), Wiley-VCH-Verlagsgemeinschaft, Weinheim.

[45] B. L. Kniep, F. Girgsdies, T. Ressler, *J. Catal.* **2005**, *236*, 34–44.

[46] T. Ressler, B. L. Kniep, I. Kasatkin, R. Schlögl, *Angew. Chem. Int. Ed. Engl.* **2005**, *44*, 4704–4707.

[47] A. Taguchi, F. Schüth, *Microporous Mesoporous Mater.* **2005**, *77*, 1–45.

[48] Z. Al Othman, *Materials* **2012**, *5*, 2874–2902.

[49] M.-J. Suh, S.-K. Ihm, *Top. Catal.* **2010**, *53*, 447–454.

[50] P. Munnik, M. Wolters, A. Gabrielsson, S. D. Pollington, G. Headdock, J. H. Bitter, P. E. de Jongh, K. P. de Jong, *J. Phys. Chem. C* **2011**, *115*, 14698–14706.

[51] J. S. Yang, W. Y. Jung, G.-D. Lee, S. S. Park, S.-S. Hong, *Top. Catal.* **2010**, *53*, 543–549.

[52] K. Cassiers, T. Linssen, M. Mathieu, M. Benjelloun, K. Schrijnemakers, P. van der Voort, P. Cool, E. F. Vansant, *Chem. Mater.* **2002**, *14*, 2317–2324.

[53] G. Prieto, J. Zečević, H. Friedrich, K. P. de Jongh, P. E. de Jongh, *Nat. Mater.* **2013**, *12*, 34–39.

[54] I. J. Drake, K. L. Fujdala, S. Baxamusa, A. T. Bell, T. D. Tilley, *J. Phys. Chem. B* **2004**, *108*, 18421–18434.

[55] G. Centi, B. Wichterlová, A. T. Bell, *Catalysis by unique metal ion structures in solid matrices. From science to application*, Kluwer Academic Publishers, Dordrecht, Boston, **2001**, *v. 13*.

[56] M. M. Günther, Dissertation, Technische Universität Berlin, Berlin, **2001**.

[57] S. Fujita, Y. Kanamori, A. M. Satriyo, N. Takezawa, *Catal. Tod.* **1998**, *45*, 241–244.

[58] D. Zhao, J. Feng, Huo, Q., Melosh, N., G. H. Fredrickson, B. F. Chmelka, G. D. Stucky, *Science* **1998**, *279*, 548–552.

[59] J. Scholz, A. Walter, T. Ressler, *J. Catal.* **2014**, *309*, 105–114.

[60] T. Ressler, *J. Synchrotron Rad.* **1998**, *5*, 118–122.

[61] P. Krawiec, C. Weidenthaler, S. Kaskel, *Chem. Mater.* **2004**, *16*, 2869–2880.

[62] a) A. Silvestre-Alberto, E. O. Jardim, E. Bruijn, V. Meynen, P. Cool, A. Sepulveda-Escribano, J. Silvestre-Alberto, F. Rodriguez-Reinoso, *Langmuir* **2009**, *25*, 939–943; b) M. A. Smith, R. F. Lobo, *Microporous Mesoporous Mater.* **2010**, *131*, 204–209;

[63] M. C. Marion, E. Garbowski, M. Primet, *Faraday Trans.* **1990**, *86*, 3027–3032.

[64] L. Levien, C. T. Prewitt, D. J. Weidner, *Am. Mineral.* **1980**, *65*, 920–930.

[65] N. J. Calos, J. S. Forrester, G. B. Schaffer, *J. Solid State Chem.* **1996**, *122*, 273–280.

[66] A. Kirfel, K. Eichhorn, *Acta Crystallogr. A Found. Crystallogr.* **1990**, *46*, 271–284.

[67] H. Praliaud, S. Mikhailenko, Z. Chajar, M. Primet, *Appl. Catal., B* **1998**, *16*, 359–374.

[68] H. A. Jahn, E. Teller, *Proceedings of the Royal Society A: Mathematical, Physical and Engineering Sciences* **1937**, *161*, 220–235.

[69] R. E. Newnham, R. P. Santoro, *Phys. Stat. Sol.* **1967**, *19*, K87-K90.

[70] Y. Itho, S. Nishiyama, S. Tsuruya, M. Masai, *J. Phys. Chem.* **1994**, *98*, 960–967.

[71] M. H. Groothaert, J. A. van Bokhoven, A. A. Battiston, B. M. Weckhuysen, R. A. Schoonheydt, *J. Am. Chem. Soc.* **2003**, *125*, 7629–7640.

[72] S. Velu, K. Suzuki, M. Okazaki, M. P. Kapoor, T. Osaki, F. Ohashi, *J. Catal.* **2000**, *194*, 373–384.

[73] Z. R. Ismagilov, S. A. Yashnik, V. F. Anufrienko, T. V. Larina, N. T. Vasenin, N. N. Bulgakov, S. V. Vosel, L. T. Tsykoza, *Appl. Surf. Sci.* **2004**, *226*, 88–93.

[74] K. D. Karlin, J. C. Hayes, Y. Gultneh, R. W. Cruse, J. W. McKown, J. P. Hutchinson, J. Zubieta, *J. Am. Chem. Soc.* **1984**, *106*, 2121–2128.

[75] H. Haug, S. W. Koch, *Quantum theory of the optical and electronic properties of semiconductors,* 3rd ed. ed., World Scientific, Singapore, **1994**.

[76] J. Liu, X. Huang, Y. Li, K. M. Sulieman, X. He, F. Sun, *Cryst. Growth Des.* **2006**, *6*, 1690–1696.

[77] F. Marabelli, G. Parravicini, F. Salghetti-Drioli, *Phys. Rev. B* **1995**, *52*, 1433–1436.

[78] R. Weber, *J. Catal.* **1995**, *151*, 470–474.

[79] H. Chen, G. Zhao, Y. Liu, *Mater. Lett.* **2013**, *93*, 60–63.

[80] J. E. Hahn, R. A. Scott, K. O. Hodgson, S. Doniach, S. R. Desjardins, E. I. Solomon, *Chem. Phys. Lett.* **1982**, *88*, 595–598.

[81] R. Bair, W. Goddard, *Phys. Rev. B* **1980**, *22*, 2767–2776.

[82] I. J. Pickering, G. N. George, *Inorg. Chem.* **1995**, *34*, 3142–3152.

[83] J. Tranquada, S. Heald, A. Moodenbaugh, *Phys. Rev. B* **1987**, *36*, 5263–5274.

[84] D. Grandjean, V. Pelipenko, E. D. Batyrev, J. C. van den Heuvel, A. A. Khassin, T. M. Yurieva, B. M. Weckhuysen, *J. Phys. Chem. C* **2011**, *115*, 20175–20191.

[85] R. W. G. Wyckoff, *Cryst. Struct.* **1963**, *1*, 85–237.

[86] A. Walter, Dissertation, Technische Universität Berlin, Berlin, **2011**.

[87] J. F. Pérez-Robles, F. J. García-Rodríguez, J. M. Yáñez-Limón, F. J. Espinoza-Beltrán, Y. V. Vorobiev, J. González-Hernández, *J. Phys. Chem. Sol.* **1999**, *60*, 1729–1736.

[88] H. H. Otto, M. Meibohm, *Z. Kristallogr.* **1999**, *214*, 558–565.

[89] N. V. Belov, B. A. Maksimov, Y. Z. Nozik, L. A. Muradyan, *Dokl. Akad. Nauk SSSR* **1978**, *239*, 842–845.

[90] H. R. Oswald, A. Reller, H. W. Schmalle, E. Dubler, *Acta Crystallogr., Sect. C: Cryst. Struct. Commun.* **1983**, *39*, 2279–2284.

[91] F. Zigan, W. Joswig, H. U. Schuster, S. A. Masons, *Z. Kristallogr.* **1977**, *145*, 412–426.

[92] J. Scholz, A. Walter, A. H. P. Hahn, T. Ressler, *Microporous and Mesoporous Materials* **2013**, *180*, 130–140.

[93] R. Zubrzycki, J. D. Epping, T. Ressler, *ChemCatChem* **2015**, *7*, 1112–1121.

[94] A. Walter, R. Herbert, C. Hess, T. Ressler, *Chemistry Central Journal* **2010**, *4*, 3.

[95] J. Florek-Milewska, P. Decyk, M. Ziolek, *Appl. Catal. A* **2011**, *393*, 215–224.

[96] G. C. Chinchen, M. S. Spencer, K. C. Waugh, D. A. Whan, *J. Chem. Soc., Faraday Trans. 1* **1987**, *83*, 2193–2212.

[97] M. M. Günter, B. Bems, R. Schlögl, T. Ressler, *J. Synchrotron Rad.* **2001**, *8*, 619–621.

[98] C. Baltes, S. Vukojevic, F. Schüth, *J. Catal.* **2008**, *258*, 334–344.

[99] S. Sakong, A. Groß, *Surf. Sci.* **2003**, *525*, 107–118.

[100] R. Hoffmann, *Solids and surfaces. A chemist's view of bonding in extended structures*, VCH Publishers, New York, NY, **1988**.

[101] B. Hammer, J. K. Nørskov, *Adv. Catal.* **2000**, *45*, 71–129.

[102] J. L. Carter, J. A. Cusumano, J. H. Sinfelt, *J. Phys. Chem.* **1966**, *70*, 2257–2263.

[103] A. L. M. da Silva, J. P. den Breejen, L. V. Mattos, J. H. Bitter, K. P. de Jong, F. B. Noronha, *J. Catal.* **2014**, *318*, 67–74.

[104] S. Polarz, F. Neues, M. W. E. van den Berg, W. Grünert, L. Khodeir, *J. Am. Chem. Soc.* **2005**, *127*, 12028–12034.

[105] R. Naumann d'Alnoncourt, M. Kurtz, H. Wilmer, E. Löffler, V. Hagen, J. Shen, M. Muhler, *J. Catal.* **2003**, *220*, 249–253.

[106] P. L. Hansen, J. B. Wagner, S. Helveg, J. R. Rostrup-Nielsen, B. S. Clausen, H. Topsøe, *Science* **2002**, *295*, 2053–2055.

[107] J. W. Evans, M. S. Wainwright, A. J. Bridgewater, D. J. Young, *Appl. Catal.* **1983**, *7*, 75–83.

[108] S. Sato, R. Takahashi, T. Sodesawa, K.-i. Yuma, Y. Obata, *J. Catal.* **2000**, *196*, 195–199.

[109] O. Hinrichsen, T. Genger, M. Muhler, *Chem. Eng. Technol.* **2000**, *23*, 956–959.

[110] B. Dvořàk, J. Pašek, *J. Catal.* **1970**, *18*, 108–114.

[111] D. A. Monti, A. Baiker, *J. Catal.* **1983**, *83*, 323–335.

[112] R. M. Dell, F. S. Stone, P. F. Tiley, *Trans. Faraday Soc.* **1953**, *49*, 195–201.

[113] R. J. Madon, M. Boudart, *Ind. Eng. Chem. Fund.* **1982**, *21*, 438–447.

[114] E. A. Owen, E. L. Yates, *Philos. Mag.* **1933**, *15*, 472–488.

[115] H. M. Rietveld, *J. Appl. Crystallogr.* **1969**, *2*, 65–71.

[116] P. Scherrer, *Nachr. Ges. Wiss. Göttingen, Math.-Phys. Kl.* **1918**, *2*, 98–100.

[117] T. H. de Keijser, E. J. Mittemeijer, H. C. F. Rozendaal, *J. Appl. Cryst.* **1983**, *16*, 309–316.

[118] T. Ressler, *Anal. Bioanal. Chem.* **2003**, *376*, 584–593.

[119] B. S. Clausen, J. K. Nørskov, *Top. Catal.* **2000**, *10*, 221–230.

[120] C. J. G. van der Grift, A. F. H. Wielers, B. P. J. Jogh, J. van Beunum, M. de Boer, M. Versluijs-Helder, *J. Catal.* **1991**, *131*, 178–189.

[121] Y. Zhang, N. Zheng, K. Wang, S. Zhang, J. Wu, *J. Nano Mat.* **2013**, *2013*, 1–7.

[122] J. Y. Carriat, M. Che, M. Kermarec, A. Decarreau, *Catal. Lett.* **1994**, *25*, 127–140.

[123] R. J. Matyi, L. H. Schwartz, J. B. Butt, *Catal. Rev.* **1987**, *29*, 41–99.

[124] a) N. Koga, J. M. Criado, *J. Am. Ceram. Soc* **1998**, *81*, 2901–2909; b) N. Koga, J. M. Criado, *J. Thermal Anal.* **1997**, *49*, 1477–1484;

[125] S. R. Bare, T. Ressler in *Advances in Catalysis* (Eds.: B. C. Gates, H. Knözinger, F. C. Jentoft), Academic Press, Amsterdam, **2009**.

[126] J. A. Rodriguez, J. Y. Kim, J. C. Hanson, M. Pérez, A. I. Frenkel, *Catal. Lett.* **2003**, *85*, 247–254.

[127] K. L. Fujdala, I. J. Drake, A. T. Bell, T. D. Tilley, *J. Am. Chem. Soc.* **2004**, *126*, 10864–10866.

[128] X. Xu, S. M. Vesecky, J. W. He, D. W. Goodman, *J. Vac. Sci. Technol. A* **1993**, *11*, 1930–1935.

[129] I. Böttger, T. Schedel-Niedrig, O. Timpe, R. Gottschall, M. Hävecker, T. Ressler, R. Schlögl, *Chem. Eur. J.* **2000**, *6*, 1870–1876.

[130] B. A. Sexton, *Surf. Sci.* **1979**, *88*, 299–318.

[131] B. A. Sexton, A. E. Hughes, N. R. Avery, *Surf. Sci.* **1985**, *155*, 366–386.

[132] B. Hammer, *Top. Catal.* **2006**, *37*, 3–16.

[133] a) J. J. Rehr, R. C. Albers, S. I. Zabinsky, *Phys. Rev. Lett.* **1992**, *69*, 3397–3400; b) A. I. Frenkel, C. W. Hills, R. G. Nuzzo, *J. Phys. Chem. B* **2001**, *105*, 12689–12703;

[134] S. I. Zabinsky, J. J. Rehr, A. Ankudinov, R. C. Albers, M. J. Eller, *Phys. Rev. B* **1995**, *52*, 2995–3009.

[135] B. S. Clausen, H. Topsøe, L. B. Hansen, P. Stoltze, J. K. Nørskov, *Jpn. J. Appl. Phys.* **1993**, *32*, 95–97.

[136] H. G. Fritsche, R. E. Benfield, *Z. Phys. D.* **1993**, *26*, 15–17.

[137] B. S. Clausen, L. Gråbæk, H. Topsøe, L. B. Hansen, P. Stoltze, J. Nørskov, O. H. Nielsen, *J. Catal.* **1993**, *141*, 368–379.

[138] G. Apai, J. F. Hamilton, J. Stohr, A. Thompson, *Phys. Rev. Lett.* **1979**, *43*, 165–169.

[139] J. Agrell, H. Birgersson, M. Boutonnet, *J. Power Source* **2002**, *106*, 249–257.

[140] H. Purnama, T. Ressler, R. Jentoft, H. Soerijanto, R. Schlögl, R. Schomäcker, *Appl. Catal., A* **2004**, *259*, 83–94.

[141] H. Purnama, F. Girgsdies, T. Ressler, J. H. Schattka, R. A. Caruso, R. Schomäcker, R. Schlögl, *Catal. Lett.* **2004**, *94*, 61–68.

[142] a) J. W. Coenen, *Appl. Catal.* **1991**, *75*, 193–223; b) Y. Matsumura, K. Kuraoka, T. Yazawa, M. Haruta, *Catal. Tod.* **1998**, *45*, 191–196;

[143] F. Girgsdies et al., *Catal. Lett.* **2005**, *102*, 91–97.

[144] X. Xu, S. M. Vesecky, D. W. Goodman, *Science* **1992**, *258*, 788–790.

[145] a) J. L. Carter, J. H. Sinfelt, *J. Phys. Chem.* **1966**, *70*, 3003–3006; b) M. Valden, X. Lai, D. W. Goodman, *Science* **1998**, *281*, 1647–1650;

[146] S. Gan, Y. Liang, D. R. Baer, M. R. Sievers, G. S. Herman, C. H. F. Peden, *J. Phys. Chem. B* **2001**, *105*, 2412–2416.

[147] R. M. Rioux, H. Song, J. D. Hoefelmeyer, P. Yang, G. A. Somorjai, *J. Phys. Chem. B* **2005**, *109*, 2192–2202.

[148] M. V. Twigg, M. S. Spencer, *Appl. Catal., A* **2001**, *212*, 161–174.

[149] B. Frank, F. C. Jentoft, H. Soerijanto, J. Kröhnert, R. Schlögl, R. Schomäcker, *J. Catal.* **2007**, *246*, 177–192.

[150] I. Fisher, A. T. Bell, *J. Catal.* **1999**, *184*, 357–376.

[151] J. S. Campbell, *Ind. Eng. Chem. Proc. Des. Dev.* **1970**, *9*, 588–595.

[152] M. S. Spencer, *Nature* **1986**, *323*, 685–687.

[153] G. Tammann, Q. A. Mansuri, *Z. Anorg. Allg. Chem.* **1923**, *126*, 119–128.

[154] E. Ruckstein, B. Pulvermacher, *J. Catal.* **1973**, *29*, 224–245.

[155] C. T. Campbell, S. C. Parker, D. E. Starr, *Science* **2002**, *298*, 811–814.

[156] A. Mirhosseini, A. Fazeli, A. Abbas, Y. Mortazavi, *Iran. J. Chem. Chem. Eng.* **2013**, *32*, 45–59.

[157] M. Argyle, C. Bartholomew, *Catalysts* **2015**, *5*, 145–269.

[158] M. M. Hauman, A. Saib, D. J. Moodley, E. du Plessis, M. Claeys, E. van Steen, *ChemCatChem* **2012**, *4*, 1411–1419.

[159] A. F. Holleman, E. Wiberg, N. Wiberg, *Lehrbuch der anorganischen Chemie,* 102., stark umgearbeitete und verb. Aufl. ed., de Gruyter, Berlin, New York, **2007**.

[160] T. Ressler, S. L. Brock, J. Wong, S. L. Suib, *J. Phys. Chem. B* **1999**, *103*, 6407–6420.

[161] C. Kaito, K. Fujita, H. Hashimoto, *Jpn. J. Appl. Phys.* **1973**, *12*, 489–496.

[162] E. O. Kirkendall, C. Upthegrove, L. Thomassen, C. Upthegrove, *Trans. AIME* **1939**, *133*, 186–203.

[163] S. J. L. Billinge, I. Levin, *Science* **2007**, *316*, 561–565.

[164] J. Le Bars, U. Specht, J. S. Bradley, D. G. Blackmond, *Langmuir* **1999**, *15*, 7621–7625.

Epilogue

Danksagung

Als erstes möchte ich mich bei den Gutachtern dieser Arbeit Prof. Ressler und Prof. Behrens sowie dem Prüfungsvorsitzenden Prof. Friedrich für Ihre Mühen bedanken. Besonders danke ich Prof. Ressler für die Überlassung des Themas und die Möglichkeit, via Quereinstieg die Geheimnisse der Katalyseforschung kennenzulernen. Ich bin ihm auch für die weitreichenden Befugnisse der Experimentgestaltung, Gerätebetreuung und Studentenbetreuung dankbar. Ebenso bedanke ich mich bei ihm für die anregenden Diskussionen, die mich stets weitergebracht haben.

Ein herzliches Dankeschön geht vor allem ans Kollegium im Arbeitskreis Ressler: für die technische Hilfe an Dr. T. C. Rödel und A. Hahn; die Exxen Dr. A. Walter, Dr. J. Scholz, Dr. R. Zubrzycki, A. Müller und Sven für Diskussionen, Kaffee und gemeinsame Strahlzeiten mit allen damit verbundenen Höhen und Tiefen; Semiha Schwarz als Gut-Laune-Quelle in C51 und als helfende Hand im Labor, S. Krombach und A. Rahmel für Bürokratiebewältigung, Zuspruch, Smalltalk und Material. Ein großes Dankeschön geht auch an die Mitarbeiter des Institutes für Chemie an der TU Berlin. Besonders danke ich dem HASYLAB bei DESY in Hamburg. Erst durch die genehmigten Messzeiten und durch finanzielle Unterstützung ist ein wesentlicher Teil dieser Arbeit möglich geworden. Dazu danke ich dem Team vor Ort: Dr. A. Webb, Dr. R. Chernikov, Dr. M. Murphy, M. Hermann und U. Brüggemann.

Wesentliche Fortschritte während dieser Arbeit erzielte ich während der Betreuung von den Studienarbeiten von L. Schmidt, R. Kaml und K. Pavel. Ein Dankeschön für Euer Mitwirken! Ebenso danke ich M. König und M. Dieckmann für die weiterführenden Untersuchungen an CuO/SBA-15 Katalysatoren.

Prof. Thomas und Dr. A. Grigas möchte ich für die Starthilfe bei der SBA-15 Synthese danken. Ohne ihre Unterstützung und ihre Ratschläge wäre eine reproduzierbare Synthese des weißen Pulvers nicht so schnell erreicht worden. Das hat mir und dem Arbeitskreis große Pforten geöffnet. Vielen Dank. Dr. G. Hörner danke ich für den einfachen und doch so fruchtbaren Einstieg in die Analyse von DR-UV/Vis-Spektren. A. Müller danke ich für den umfassenden Überblick über Reflexanpassungen.

Prof. Behrens möchte ausdrücklich für die fruchtbare und angenehme Zusammenarbeit bei anderen Projekten danken. Mein Dank gilt auch seinen Mitarbeitern, Dr. S. Zander, Dr. S. Kühl und Herrn M. Jastak.

Für stetes Voranbringen als externe Motivatoren möchte ich mich ausdrücklich bedanken bei: Dr. A. Walter, R. Schenk, Dr. K. Hahnewald, Dr. N. Nischan, Dr. D. Köhler, N. Dietrich.

Vielen Dank für alles und die begleitende Unterstützung an meine Eltern! Danke Marlen, für Deinen Zuspruch, Deine Geduld und die schönen Zerstreuungen.

„Das Problem ist immer, mit der Arbeit fertig zu werden, in dem Gedanken, nie und mit nichts fertig zu werden . . ., es ist die Frage: weiter, rücksichtslos weiter, oder aufhören, schlußmachen . . . es ist Frage des Zweifels, des Mißtrauens und der Ungeduld."

Thomas Bernhard in *Meine Preise*